GLASS:
SCIENCE AND TECHNOLOGY

VOLUME 5
Elasticity and Strength in Glasses

Contributors

A. S. ARGON
ROGER F. BARTHOLOMEW
F. M. ERNSBERGER
S. W. FREIMAN
ROBERT GARDON
HARMON M. GARFINKEL

GLASS:
SCIENCE AND TECHNOLOGY

Edited by   *D. R. UHLMANN*

DEPARTMENT OF MATERIALS SCIENCE
AND ENGINEERING
MASSACHUSETTS INSTITUTE OF TECHNOLOGY
CAMBRIDGE, MASSACHUSETTS

N. J. KREIDL

DEPARTMENT OF CHEMICAL
AND NUCLEAR ENGINEERING
UNIVERSITY OF NEW MEXICO
ALBUQUERQUE, NEW MEXICO

VOLUME 5
Elasticity and Strength in Glasses

 1980

ACADEMIC PRESS
A Subsidiary of Harcourt Brace Jovanovich, Publishers

New York   London   Toronto   Sydney   San Francisco

ACADEMIC PRESS, INC.
111 Fifth Avenue, New York, New York 10003

United Kingdom Edition published by
ACADEMIC PRESS, INC. (LONDON) LTD.
24/28 Oval Road, London NW1 7DX

Library of Congress Cataloging in Publication Data
Main entry under title:

Elasticity and strength in glasses.

(Glass ; v. 5)
Includes index.
1. Glass. 2. Elasticity. I. Uhlmann, Donald
Robert. II. Kreidl, N. J. III. Series.
TP848.G56 vol. 5 [TA450] 666'.1s [620.1'4432] 80−38
ISBN 0−12−706705−1 (v. 5)

PRINTED IN THE UNITED STATES OF AMERICA

80 81 82 83   9 8 7 6 5 4 3 2 1

Contents

List of Contributors

Numbers in parentheses indicate the pages on which the authors' contributions begin.

A. S. ARGON (79), *Department of Mechanical Engineering, Massachusetts Institute of Technology, Cambridge, Massachusetts 02139*

ROGER F. BARTHOLOMEW (217), *Corning Glass Works, Sullivan Park, Corning, New York 14830*

F. M. ERNSBERGER (1, 133), *PPG Industries, Inc., Pittsburgh, Pennsylvania 15238*

S. W. FREIMAN (21), *Fracture and Deformation Division, National Bureau of Standards, Washington, D.C. 20234*

ROBERT GARDON (145), *Engineering and Research Staff, Ford Motor Company, Dearborn, Michigan 48121*

HARMON M. GARFINKEL (217), *Corning Glass Works, Sullivan Park, Corning, New York 14830*

Preface

Recent years have seen a notable series of developments that have emphasized the importance of amorphous solids (glasses). These include activity in the areas of fiber optics, optical waveguides, amorphous semiconductors, glass lasers, glass–ceramic materials, and photochromic glasses, to name only a few. In addition to advances related to these areas of technological application, there have also been notable improvements in our understanding of the structure, processing, and properties of glass-forming materials. Regrettably, most of the information developed in these areas is contained in a myriad of individual research papers and in-house knowledge, rather than in any convenient central reference source. The time seemed appropriate, therefore, to bring together the available knowledge about this important class of materials.

"Glass: Science and Technology," of which this volume is the first to appear, will be broad in its scope. The topics covered will range from fundamental understanding of the structure and properties of glasses—based on applying the disciplines of physics, chemistry, and materials science to amorphous materials—to highly applied areas such as the melting of glass, its forming in various shapes, and the use of glass in various important technological applications. Wherever possible, the common features and notable differences between glasses and other types of materials will be highlighted.

At this time 12 volumes are planned. The tentative program is

1	Structure of Glass
2	Submicrostructure of Glass
3	Glass Forming Systems and Glass Ceramic Materials
4	Diffusion, Viscous Flow, and Relaxation Phenomena
5	Elasticity and Strength in Glasses
6	Electrical Properties of Glasses
7	Optical Properties of Glasses
8	Processing of Oxide Glasses
9	Processing of Polymers
10	Thermal and Magnetic Properties of Glasses
11	Glass Surfaces
12	Tabulation of Data

In view of the rapid progress in understanding and controlling mechanical properties during the past years it seemed appropriate to come out with Volume 5 first. It appears that much of the confusion about competitive views of glass as brittle or deformable can now be resolved by relatively simple concepts.

Engineers have developed and successfully used ideas that clearly distinguish the mechanical properties of the glass from the influence of the number, distribution, size, and shape of flaws, particularly in the surface. Fracture mechanics has been mobilized to predict strength and lifetime under defined service conditons. The classic art of increasing surface strength by compression in the flawed surface layer, using controlled quenching, has been subjected to careful analysis leading to improved products. At the same time numerous new methods of introducing compression layers by manipulating composition and structure have become available and are understood in considerable detail. The editors are confident that in this treatise the dissemination of the most recent knowledge and its use in this field has been entrusted to highly competent workers.

GLASS:
SCIENCE AND TECHNOLOGY

VOLUME 5
Elasticity and Strength in Glasses

CHAPTER 1

Elastic Properties of Glasses

F. M. Ernsberger

PPG INDUSTRIES, INC.
PITTSBURGH, PENNSYLVANIA

I. Introduction

This review is limited to vitreous silica and the silicate glasses, including those that contain moderate amounts of other network formers such as B_2O_3, Al_2O_3, and P_2O_5. Chalcogenide, metallic, and organic glasses are specifically excluded, although certain concepts may be applicable to these glasses as well.

It is assumed that the reader is familiar with the definitions and interrelationships of the three constants used in small-deformation stress–strain analysis of isotropic materials. With minor exceptions, glasses are fully isotropic, even in the form of drawn fibers.

No attempt has been made to assemble data as such; the treatment is meant instead to identify concepts that have general validity and to emphasize recent developments. The older literature, including much systematic

1

property data, is reviewed in a treatise by Morey (1954). The subject is also briefly reviewed in a recent treatise by Babcock (1977).

In the final two sections, the discussion will depart from the domain of true elasticity to provide an introduction to the interesting and controversial behavior of glass at very high levels of stress, such as those applied in an indentation hardness test. Under these conditions, brittle glasses behave in what looks very much like a plastic manner.

II. The Nature of Elasticity in Glasses

There is a widespread opinion that glasses are supercooled liquids and therefore have a finite viscosity at ordinary ambient temperatures. Stories are told of glasses flowing under their own weight: of ancient windowpanes that are thicker at the bottom; of glass that has sagged in storage. These observations must find other explanations, because glasses of commercially useful compositions are in fact rigid solids at ordinary temperatures. Vitreous silica in particular is a nearly ideal elastic material; that is, it does not creep under load, and it recovers instantly after a prolonged deformation.

On the other hand, glasses that contain substantial amounts of network-modifying oxides (Na_2O, CaO, etc.) often exhibit both creep and delayed recovery. It is probably this behavior that is responsible for many of the observations that have been erroneously ascribed to cold viscous flow. These delayed-elastic effects have such long relaxation times that they are sometimes difficult to distinguish from true viscous flow. Douglas (1966) stated that "any experiment which purports to measure a viscosity greater than 10^{16} poises needs to be examined extremely carefully."

The atomic mechanisms responsible for these ambient-temperature viscoelastic effects have not been fully explicated, but there is little doubt that monovalent cations are involved. The mobility of these ions is severely restricted by steric and electrostatic considerations, but nevertheless internal-friction and conductivity measurements show that cation mobility is detectable down to temperatures approaching that of liquid nitrogen.

The time-dependent nature of the stress–strain behavior of glasses naturally becomes more pronounced as the temperature is raised. These effects constitute a specialized technical study that will be discussed under the general topic "Relaxation Phenomena" in this volume. For present purposes it is enough to state that the time dependence of mechanical behavior of the glasses under consideration is seldom large enough to influence the measurement of elastic constants, particularly at the high measurement frequencies normally employed.

A more-significant consideration, and one peculiar to glasses, is the state of anneal, sometimes termed "stabilization." The elastic constants, along with density, refractive index, and other intensive properties, change significantly with the rate of cooling of a glass through its "transformation range." The transformation range of a glass may be defined as that temperature range within which its properties spontaneously change at an experimentally observable rate. The property changes arise from changes in the density of packing. Obviously the limits of this range have no fundamental definition, but are determined by the time allotted to the experiment and by the sensitivity of the measurement. In practical terms, the existence of this stabilization effect means that property data of extraordinary precision have no particular absolute significance unless the thermal history of the sample is carefully specified.

III. Measurement of Elastic Constants

Ultrasonic techniques continue to be favored for measurement of the elastic constants of glasses. Manghnani and his co-workers specialize in this field. They have elevated the pulse–echo technique to new levels of precision and accuracy by the superposition of pulses and echoes in the manner of interferometry. Even the elusive Poisson ratio is being reported to four significant figures.

With such precision at hand, it becomes profitable to study the effects of pressure and temperature on the elastic constants. Manghnani et al. (1969) measured the constants for Vycor high-silica glass to 8 kbar. Skolowski and Manghnani (1969) measured calcium aluminate glasses to 3.5 kbar. Subsequently, Manghnani (1972) measured six glasses in the $Na_2O–TiO_2–SiO_2$ system to 7 kbar and 300°. Manghnani (1974) has also studied the new low-expansion glasses in the $SiO_2–TiO_2$ system as a function of temperature, pressure, and composition. Data on a commercially produced soda-lime glass to 400° are awaiting publication.

An elegant new measurement technique appeared recently with the publication of a paper by Huang et al. (1973) on a determination of elastic constants by Brillouin scattering. A few words about the principles underlying this unusual technique are appropriate.

At any temperature above absolute zero, all elastic solids contain thermally activated density fluctuations that propagate within the solid in all directions at sonic velocities with many superimposed frequencies. If a transparent elastic solid is probed with a laser, Brillouin scattering occurs from those trains of density waves whose frequency and direction of propagation happen to satisfy the Bragg condition for the laser beam; Rayleigh scattering occurs as well, of course. Thus the spectrum of the scattered light consists of five lines: a central Rayleigh peak at the laser frequency flanked symmetrically by pairs of sum-and-difference peaks. One pair reveals the frequency of the longitudinal acoustic mode; the other, that of the transverse mode. Thus it is possible to calculate both Young's modulus and the shear modulus and, from these, Poisson's ratio.

The acoustic frequencies involved are in the hypersonic range; that is, in gigahertz. Nevertheless the elastic constants are not distinguishably different from those measured at low frequencies.

An unusual ultrasonic technique was described by Fraser and LeCraw (1964). The elastic constants were deduced from observations of resonances excited in a small *spherical* specimen. Soga and others (1967, 1968) subsequently applied this technique in several interesting studies of glasses and ceramics. The main advantage seems to be in the small size (300 μm–5 mm) and simple geometry of the samples that are used.

A novel method for the measurement of elastic constants was demonstrated by Sinha (1977). A strip of elastic material whose thickness is not too small relative to its width will exhibit a saddle-shaped (anticlastic) curvature when bent. The elastic constants can be deduced from appropriate measurements of this complex curvature. The results are not highly precise (only two significant figures were obtained), but the method has the advantage that it is applicable at elevated sample temperatures when noncontacting optical methods are used to obtain the curvatures.

Elastic constants can sometimes be estimated from information that might seem to be totally unrelated. For example, Szigeti (1950) proposed an equation relating the bulk modulus of crystals to the maximum frequency of infrared reflection. Anderson (1965) subsequently demonstrated the applicability of this equation to glass.

The existence of this relationship illustrates the great fundamental significance of elastic constants, related as they are to interatomic forces and vibrational frequencies. Sanditov and Bartenev (1973) have proposed relationships between bulk modulus and still other solid-state properties such as thermal expansion coefficient, glass transition temperature, and microhardness. It does not appear that these have yet been independently verified.

IV. Prediction of Elastic Constants

One of the ultimate goals of scientific endeavor is to make empirical measurements unnecessary. This goal has been attained when our understanding of a phenomenon becomes so profound that the desired quantities can be calculated from a few fundamental constants. A paper by Makishima and Mackenzie (1973) embodies a significant degree of progress in this direction.

The derivation is adapted from an existing theoretical treatment of

Young's modulus of ionic crystals. In this treatment a coulomb interionic potential is assumed, corrected for the overlapping potentials of neighboring ionic fields by introduction of the Madelung constant. The resulting formulation cannot be directly applied to glasses because of the difficulty of evaluating the Madelung constant of a disordered structure. This difficulty was bypassed by noting that the term involving the Madelung constant is a binding energy per unit volume. For glasses, this quantity is expressible as a product of the dissociation energy per unit volume and the packing density of ions. Both can be evaluated from the composition and density of the glass, and the molar dissociation energy and molar volume of each component oxide. These are the only empirical data required.

The relationship was tested by comparison with Young's modulus data for 30 glasses containing all the common glass-making oxides except B_2O_3. The calculated modulus seems to have a systematic tendency to be low by about 5%. Nevertheless this is surely a creditable accomplishment, considering the wide range of glass types and the fundamental nature of the input data.

For borate glasses, the simultaneous presence of both 3- and 4-coordinated boron must be taken into consideration by using dissociation energies appropriate to each type of coordination. An empirically determined fraction expressing the proportion of the two states is also necessary. Phase separation, if extensive, probably invalidates the computation. Borosilicate glasses were not considered.

A different approach was taken by Yamane and Sakaino (1974) in which the melting points of the constituent oxides appear as a measure of the cohesive forces in the structure:

$$E = 9.3(\rho/M) \sum_i T_{mi} X_i,$$

where E is Young's modulus in kilobars, ρ the glass density, M its mean molecular weight, T_{mi} the absolute melting point of the ith oxide, and X_i the mole fraction of that oxide. The factor 9.3 is empirical, determined by fitting the equation to the known elastic moduli of the common glass-forming oxides B_2O_3, GeO_2, and SiO_2.

V. Empirical Correlations

It has been known for many years that certain glass properties, including Young's modulus, are related in an approximately additive manner to composition. By consideration of a large amount of data on composition and property, it is possible to derive a best-fit set of coefficients that can be used to interpolate or extrapolate to compositions on which no data exist.

An extensive and useful correlation of this kind was made by Phillips (1964). The correlation was extended to alkali-free and high-modulus glasses by Williams and Scott (1970).

Ray (1971) proposed a correlation that is somewhat less empirical. He demonstrated for silicate and phosphate glasses that Young's modulus increases in approximate direct proportion to the packing density of oxygen ions. The packing density can be calculated in a straightforward manner from the composition and the observed glass density. Thus, in either of these systems, it is possible to predict the modulus by multiplying the known modulus of the parent oxide glass by a dimensionless factor expressing the packing density of oxygen in the unknown glass relative to that of the parent glass.

Soga and Anderson (1965) also noted the extraordinary sensitivity of elastic modulus to packing density. They were able to derive a theoretical basis for this relationship, beginning with the Born equation for the potential between two charged particles. Logarithmic plots of both bulk and shear moduli against volume per cation–anion pair for 29 silicate glasses and 11 crystalline oxides gave straight lines with a slope of (minus) 4 (Fig.

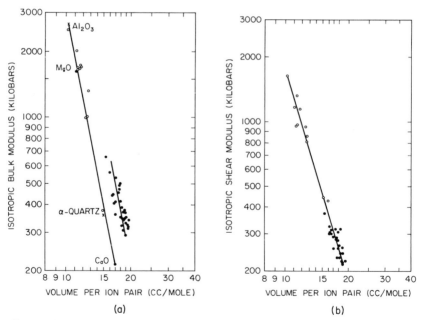

Fig. 1. Comparison of the elastic moduli of glasses and crystalline oxides. (a) ○, crystalline oxides from single crystal data; ×, crystalline oxides from Bridgman's data; ●, glass. (b) ○, crystalline oxides from single crystal data; ●, glass.

1). The moduli of oxides are much more sensitive to packing volume than are the moduli of alkali halides or other purely ionic materials, where the exponent is only $\frac{4}{3}$. The fact that glassy and crystalline oxides have the same fourth-power volume dependence indicates that interatomic potentials are largely independent of long-range order in oxide systems.

In case of the bulk modulus, the plotted lines for crystalline and glassy oxides were not coincident, however (Fig. 1a). For a given bulk modulus, glasses were found to have almost 20% more volume than crystals. In other words, the "free volume" that is characteristic of randomly packed structures has little effect on bulk modulus (except at very high pressures, as we shall see below). On the other hand, the shear-modulus curve is an extrapolation of that for crystals (Fig. 1b). We may rationalize these results on an atomistic basis by noting that volumetric changes must occur largely by bond stretching, whereas elastic shear can be accomplished by bond bending alone if there is enough room in the structure.

The temperature coefficient of the shear modulus is a matter of intense interest to the makers of ultrasonic delay devices, because the performance of these devices needs to be independent of temperature. These studies typically have a very limited scope in terms of composition, but the data are extremely precise (Inamura, 1969; Postnikov et al., 1971).

VI. The Acoustoelastic Effect

It is well known to every scientist that the velocity of propagation of light in glass is influenced by the existence of elastic strains that have a component perpendicular to the direction of propagation of the light. A publication by Meitzler and Fitch (1969) reveals that an analogous effect exists for shear-mode acoustical waves. The relative retardation of waves vibrating perpendicular and parallel to the stress vector is a linear function of stress as in the optical case, and is determined by one of the three third-order elastic constants that characterize an isotropic solid: C_{456} in the usual notation. This constant can have either sign, and accordingly the retardation can be either positive or negative.

The effect offers a way to determine C_{456}, which is a measure of the degree of anharmonicity of the interatomic vibrations, that is, of the nonlinearity of the interatomic forces with respect to atomic displacement. As a method to measure stress, it is not as sensitive as the optical effect unless the frequency of the ultrasound surpasses 2.8 GHz. This frequency is beyond the range of current technology. For opaque materials, where optical methods are inapplicable, the effect should be useful for measurement of internal stresses in spite of the sensitivity handicap.

VII. Phase Separation

The past decade has seen a rapidly growing awareness of the possibility that fine-scale separation of immiscible glassy phases may exist in systems that had been assumed to be single phase. This in turn has forced a reexamination of the variation of properties with composition, looking for the effects of phase separation. The elastic properties are among those that are being reexamined.

It is a classic problem to calculate the elastic properties of a multiphase system from the proportions and properties of the phases. Exact analytical solutions of general cases have not been obtained and probably never will be, because details of the morphology that defy exact specification have an influence on the properties of the composite. However, upper and lower bounds can be set on the possible variability by consideration of two extreme special cases. The first assumes that the material is arranged in layers parallel to the applied uniaxial stress (Voigt model). The second assumes a perpendicular arrangement (Reuss model). The Voigt model is obviously a constant-strain case, and the Reuss model, constant stress. The solution for Young's modulus is evident by inspection in the Voigt model: the resultant is simply the weighted mean of the moduli. The Reuss case is only slightly more complex.

Narrower bounds have been set by Hashin and Shtrikman (1963) for the case of spherical inclusions of one phase in a matrix of another. This configuration is experimentally realizable in certain systems such as PbO–B_2O_3. Recent measurements by Shaw and Uhlmann (1971) in this system have shown that shear, bulk, and Young's moduli are all in tolerably good agreement with the predictions of the models, and that they clearly tend to favor the lower bound of the Hashin–Shtrikman relation. Data (for Young's modulus only) are shown in Fig. 2.

In certain systems phase separation develops slowly, so that it is possible

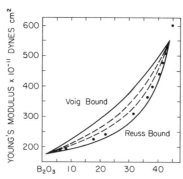

FIG. 2. Young's modulus in a two-phase system. ----, Hashin–Shtrikman bounds.

to study the effect of phase separation per se, that is, at constant composition. Zarzycki (1974) carried out this type of experiment with three potassium borosilicate glasses. He found that Young's modulus decreased by about 3% as the texture of the structure coarsened from 400 to 1500 Å. This is a significant effect, but quite small when compared with the 10^5-fold change in *viscosity* that can occur during phase separation (Simmons *et al.*, 1970).

There are several published studies of mechanical properties as a function of composition in phase-separating systems, but in no case are there discontinuities or other clear indications of the boundaries of the two-phase region. Nemilov (1972) reported Poisson's ratio to be the property most sensitive to structural changes in phase-separating glasses, but in view of the difficulty in making precise measurements of this ratio, the conclusion is open to question.

VIII. Anomalous Elastic Properties of Glasses

Vitreous silica and certain other glasses are anomalous in many respects, including some of their mechanical properties. These anomalies can in most cases be plausibly explained as consequences of the existence of "free volume" within the structure of the material. This free volume is associated with the inherent inefficiency of random packing, but more especially with the inability of the bulky tetrahedral SiO_4 groups to pack as closely together as do the spherical atoms of metals and alloys. The attainment of optimum packing is particularly hindered by the extremely high viscosity that is typical of glasses.

A. COMPRESSIBILITY

Bridgman and Simon (1953) reported that silica and borosilicate glasses increase in compressibility with increasing pressure, contrary to the behavior of other materials. The higher the silica content, the more pronounced this anomalous behavior. At a certain critical pressure the compressibility was observed to become quite large, and a portion of the compression was found to be irreversible; that is, a permanent densification occurred. This phenomenon has been studied in detail by others, notably by Mackenzie (1963) and by Cohen and Roy (1965). Poch (1967) and Mizouchi and Cooper (1971) observed the same effect with vitreous B_2O_3.

Both the anomalous compressibility and the ultimate structural collapse relate basically to the same cause, namely, inefficient packing. The unoccupied space remains potentially available, and the application of hydrostatic pressure forces the small shifts of atomic position and flexures of bonds that permit an overall decrease in the volume of the system. Much of this decrease would have occurred spontaneously had the material not had such a high viscosity. However, even low-viscosity systems that have an open structure also tend to exhibit anomalous behavior under high pressure; ordinary water is a case in point.

The degree of densification varies with pressure, temperature, time, shear stress, and possibly other variables. Exact relationships have not been established. In the case of silica, the pressure required to initiate densification is of the order of 70 kbar (1,000,000 psi) at ordinary temperatures. This pressure is too high for strictly hydrostatic experimentation. At higher temperatures, less pressure is required. In the case of B_2O_3 glass, the required pressure is lower; the effect begins at about 22 kbar at ordinary temperatures.

B. Non-Hookean Behavior

The nature of interatomic forces is such that the force constant decreases with increasing interatomic distance. Stated in macroscopic terms, Young's modulus should decrease with increasing strain. This decrease is in fact observed in the case of metallic whiskers (Brenner, 1956). Hillig (1962) discovered that the modulus of fused silica *increases* with increasing strain. Mallinder and Proctor (1964), in a more-elaborate set of measurements, found an apparently quadratic relation between stress and strain, which leads to a linear dependence of modulus upon strain:

$$E = E_0 (1 + 5.75 \, \epsilon).$$

Mallinder and Proctor (1964) highlighted the anomalous behavior of fused silica by showing that soda-lime glass decreases in modulus with increasing strain, in the manner of a "normal" material:

$$E = E_0 (1 - 5.11 \, \epsilon).$$

The explanation advanced by Mallinder and Proctor (1964) seems likely to be basically correct. Again, free volume in the silica network is held to be responsible for unusual mechanical behavior. Free volume permits the network to deform in the manner of a pantograph (Fig. 3), that is, by small

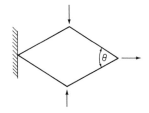

FIG. 3. Pantograph mode of deformation.

distortions of bond angles rather than by pure stretching of bonds. As strain increases (corresponding to smaller values of θ), an increasing proportion of the stress is applied to the bond-stretching mode, with a consequent increase in Young's modulus.

This concept also accounts for the surprisingly small magnitude of the elastic modulus of fused silica. Denser forms of silica should have higher values of Young's modulus, and in fact they do. The modulus for the "stiffest" direction in low quartz is nearly twice that of fused silica.

Other solids whose structures include much free volume should show similar elastic anomalies. Data are scarce, owing perhaps to the difficulty of preparing specimens that are strong enough to sustain the relatively large extensions necessary to make the effect observable. However, fibers of glassy carbon have both the necessary strength and the free-volume structure. Curtis *et al.* (1968) showed that carbon fibers have a distinctly nonlinear stress–strain behavior: Young's modulus increases as much as 30% at a strain of only 0.4%.

C. STRAIN DEPENDENCE OF POISSON'S RATIO

Mallinder and Proctor (1964) included in their experiments the measurement of rigidity modulus, and found that it also has an anomalous strain dependence. Combining the measured values of Young's modulus and rigidity modulus at corresponding strains, they calculated Poisson's ratio as a function of strain. A regular and substantial increase was observed: from 0.14 at zero strain to 0.40 at about 12% strain.

This behavior of Poisson's ratio is also consistent with the pantograph mode of deformation. This is evident from a consideration of Poisson's ratio for a simple mechanical system of jointed rigid rods. As the angle θ (Fig. 3) decreases from 180° to 0°, Poisson's ratio increases from zero to infinity.

D. TEMPERATURE DEPENDENCE OF ELASTICITY

As mentioned in Section I, a rise in temperature tends to introduce time dependence into the stress–strain behavior of most glasses. However, it is also true that temperature has an effect on the "real" or in-phase component of the complex compliance. Data of this kind are given in references cited in Section III.

Reasoning from the known character of interatomic forces, one would expect that elastic moduli would decrease with increasing temperature for any material. Metals and ceramics do generally show this behavior, but fused silica is again an exception. Sosman (1927) recorded that the torsional modulus increases with temperature at the rate of about 0.012%/°C. Again the free-volume concept offers an explanation: bonds deformed in flexure (as in the pantograph mode of compliance) would be expected to vibrate in a strongly anharmonic manner that will tend to restore the mean bond angle to its undistorted value.

An alternative explanation for the temperature-coefficient anomaly was suggested by Bozhko and Ratobylskaya (1976). Rubbery materials also exhibit a positive temperature coefficient of elastic modulus. Elastic behavior in such materials is known to be an entropy effect, so these authors suggest that there is an entropic component in the elasticity of fused silica. There appears to be little merit in this suggestion, requiring as it does the assumption of a freely hinged or freely rotating Si–OSi bond, for which there is no other evidence. To establish the existence of an entropy component in silica elasticity, it would be necessary to demonstrate that heat is evolved when silica is deformed and absorbed when it relaxes.

The addition of fluxing oxides (MO, M_2O) to silica eliminates the anomaly, and commercial silicate glasses behave in the "normal" manner. It follows that there should be intermediate compositions with a zero temperature coefficient of elastic modulus. Postnikov *et al.* (1971) reported that in the binary soda-silica system, the coefficient goes to zero at 7.5% Na_2O. Unfortunately, a glass of this composition devitrifies so easily that it is not practically useful.

IX. Hardness

The "hardness" of a material is a concept that is easily grasped intuitively, but it has proved to be difficult to define in a scientifically meaningful way. Nevertheless, in the case of metals the concept is technologically very useful, and several satisfactory measurement techniques exist. The familiar square–pyramid Vickers identation test may be the most widely used. The various methods have in common the fact that they measure the resistance of a metal to plastic deformation.

It was inevitable that someone would try the identation test on glass, although this is a brittle material for which the concept of plastic deformation seems eminently unsuitable. The first paper was published by Taylor (1949). To the surprise of everyone, permanent plastic-appearing indentations were formed. A certain amount of cracking occurred as well.

Since 1949, at least 200 papers have appeared on various aspects of the hardness of glass. The wide availability of the apparatus and the speed and simplicity of the basic measurement are surely responsible in large part for the great popularity of this field.

But popularity does not necessarily translate into fruitfulness. In this reviewer's opinion, hardness testing of glass has not been particularly fruitful. It is easy to produce data, but not so easy to establish what they mean.

An important part of the problem is the poor reproducibility of the measurement. Table I presents a survey, not necessarily exhaustive, of reported hardness values of one glass only, vitreous silica. Silica was chosen for the survey because it is a one-component glass, its surface properties are stable, and it is usually included in experimental programs as a reference substance.

TABLE I

Reported Hardness of Silica Glass

Source	Load (gm)	Hardness	Reference
G.E. fused quartz	100	473–593[a]	ASTM (1973)
Fused Brazilian quartz	100	590	Westbrook (1960)
Corning 7943	100	900	Westbrook (1960)
Unspecified	30–100	710	Ainsworth (1954)
Unspecified	50	620	Prod'homme (1968)
Unspecified	300	750	Prod'homme (1968)
Unspecified	50–500	1333[a]	Grosskopf (1969)
Corning 7940	400–1000	635	Neely (1969)
Fused Brazilian quartz	400–1000	635	Neely (1969)
G.E. fused quartz (101)	400–1000	635	Neely (1969)
Gen. Tech. fused quartz	400–1000	635	Neely (1969)

[a]Knoop indenter; remaining data by Vickers indenter.

Vitreous silica was one of four glasses used in a formal round-robin reproducibility test conducted by Committee C-14 of the American Society for Testing Materials in 1968 as one of the preliminaries to the issuance of a standardized test procedure in 1973. Six laboratories took part. Their data for fused silica at the recommended 100 gm load varied from 473 to 593 kg/mm^2 with a mean of 539. This is a spread of 22% among a select group of laboratories using a carefully standardized procedure.

Something of the magnitude of the variability when standardized procedures are *not* used may be inferred from the rest of the table. The reported hardness ranges to 1333, nearly 2.5 times the ASTM mean. Oddly enough, not one of the other laboratories reported values as low as the ASTM mean!

Parenthetically, there can be an objection to the inclusion of both Knoop and Vickers data in Table I. The shape of the two types of indenters is quite different, and there is a 7% systematic difference in the method of calculation. However, other sources of imprecision appear to be much more serious.

Some of the sources of imprecision were identified by Neely (1969). Vibrationless mounting of the test machine was found to be particularly important. However, most of the disagreement between laboratories probably arises from the difficulty of precisely locating the cusp-shaped corners of the indentations. The character and quality of the lighting and the judgment of the operator are heavily involved, and of course these are highly variable between laboratories.

Through systematic control of the many subtle variables in the hardness measurement, Neely obtained what is probably the most precise set of data in existence on the hardness of fused silica. Statistical analysis gave a distribution with 2 σ limits only 2.4% removed from the mean value; however, this mean is 18% *higher* than the mean obtained by ASTM. If we correct the data to a projected-area basis to facilitate comparison with the ASTM Knoop data, the spread *increases* by 7%.

Two conclusions are evident: first, that hardness data have only a limited absolute significance; and second, that the measurement itself is highly subjective. The first conclusion counsels caution in comparing data generated in different laboratories; the second suggests that even within a single laboratory, the data may unconsciously be manipulated by the experimenter, unless double-blind procedures are followed.

X. Microplasticity

It is possible to question the precision and accuracy of hardness measurements in the case of glassy materials, but it is not possible to deny that something interesting and anomalous occurs under an indenter. The mechanisms of plastic flow in metals, ionic crystals, and amorphous polymers are well understood. In each of these materials there is a low-energy bond type that is capable of thermal activation. The breakage of these bonds, followed by formation of new bonds with new partners in a coordinated way, makes plastic shear deformation possible. This kind of mechanism would not be expected to operate in the siloxane network, however, because the bonding is too strong. Thermal activation cannot occur until the temperature approaches the transition range for the glass. It has been suggested by Douglas (1958) that whenever a shear stress exists that is about 10% of the fundamental bond strength, then thermal activation can supply the balance of the required energy, and viscous flow will occur. This suggestion remains unconfirmed.

It has been shown, however, that another mechanism, one unique to silicate glasses, is primarily responsible for the so-called microplastic effects. Shand (1958) asserted that microindentations combine the effects of elasticity, plastic flow, and densification. Subsequent work by independent groups has abundantly verified the role of densification (Neely and Mackenzie, 1968; Ernsberger, 1968). It remains debatable, however, whether or not plastic flow plays any part.

Important new evidence in support of plastic flow has been supplied by studies with the scanning electron microscope carried out by Dick (1970) and by Peter (1970). Using sharp diamond points to probe and score glasses, they were able to demonstrate the existence of coherent spiral turnings, buttery-looking grooves, and blocky extrusions. The eye is convinced; only the mind can retain reservations.

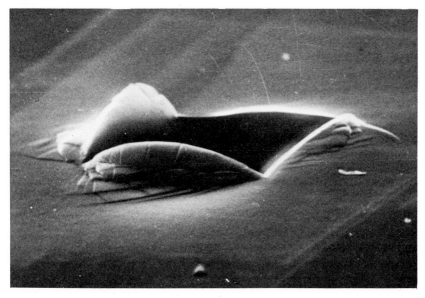

FIG. 4. Failure of glass by inhomogeneous shear. Composition: 76% SiO_2, 14% Na_2O, 10% CaO. 7000×.

A particularly interesting Peter photograph is reproduced in Fig. 4. The surface markings suggest that plastic flow in this glass occurs inhomogeneously, that is, in selected planar regions. This type of shear is already familiar in polycrystalline metals (Lüder bands). Apparently it is common in amorphous materials as well: Figure 5 shows inhomogeneous shear in an organic glass (Argon *et al.*, 1968), and Fig. 6 in a metallic glass (Gilman, 1975). However, it is significant that Peter was unable to produce plastic flow in fused silica.

If we accept these two facts, that plastic shear of inorganic glasses occurs inhomogeneously and requires the presence of network modifiers, it is possible to construct an acceptable working hypothesis for the atomistic mechanism of such flow. According to this hypothesis, the shear plane avoids the covalently bonded polysiloxane network by finding a path through the ionically bound regions of the glass. This mechanism harmonizes particularly well with the polyanion theory of glass structure, according to which silicate glasses consist of silica-rich polyanionic islands with ionically

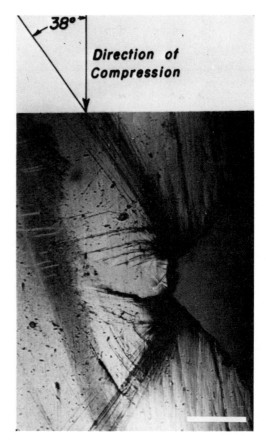

FIG. 5. Failure of polystyrene by inhomogeneous shear. White bar is 0.2 mm.

bonded interfaces. The propagation of shear in ionically bonded materials such as magnesium oxide and the alkali halides is of course well known.

To account for the inhomogeneous nature of plastic flow in amorphous materials it is necessary to postulate nucleation sites of some kind. This requires no new assumptions, for any randomly packed amorphous material will contain random local fluctuations in density and/or composition that could serve as weak points. The model proposed by Gilman (1975) for plastic flow in metallic glasses can be summarized as follows: when the shear strain in the material reaches a critical value, a portion of the strain energy is locally converted into a dilation of the structure that originates at a point of weakness and spreads rapidly in the shear plane. Dilation facilitates slip. Such a slipped and dilated region may extend entirely across the thickness of the sample and be of the order of 1 μm thick, with 10 μm of shear displacement.

FIG. 6. Failure of glassy metal by inhomogeneous shear.

Evidently, a substantial amount of energy is liberated in the limited volume of this sheared region. Part of this energy goes into the creation of the dilation that enables the atoms to glide by each other, but a part must appear as heat. The resulting local temperature rise may help to account for the occurrence of inhomogeneous shear in these glasses, and in silicate glasses as well.

Recent reviews on the subject of microhardness are available in English (Mackenzie, 1972), French (Fayet, 1969), and Spanish (Robredo *et al.*, 1970).

References

Ainsworth, L. (1954). *J. Soc. Glass Technol.* **38**, 479–547.
Anderson, O. L. (1965). *Proc. Int. Conf. Phys. Glass.* North-Holland Publ., Amsterdam.
Argon, A. S., Andrews, R. D., Godrick, J. A., and Whitney, W. (1968). *J. Appl. Phys.* **39**, 1899–1906.
ASTM (1973). These data, first published in 1973, are reprinted in the current (1978) *Annu. ASTM Stand.* Part 17, p. 750.
Babcock, C. L. (1977). *"Silicate Glass Technology Methods,"* Chapter 13. Wiley, New York.

Bozhko, Yu. A., and Ratobylskaya, V. A. (1976). *Fiz. Khim. Stekla.* **2**, 323–327 (English trans.: *Sov. J. Glass Phys. Chem.* **2**, 318–321.)

Brenner, S. S. (1956). *J. Appl. Phys.* **27**, 1484–1491.

Bridgman, P. W., and Simon, I. (1953). *J. Appl. Phys.* **24**, 405–413.

Cohen, H. M., and Roy, R. (1965). *Phys. Chem. Glasses* **6**, 149–161.

Curtis, G. J., Milne, J. M., and Reynolds, W. N. (1968). *Nature (London)* **220**, 1024–1025.

Dick, E. (1970). *Glastech. Ber.* **43** (1), 16–21.

Douglas, R. W. (1958). *J. Soc. Glass Technol.* **42** (206), 145–57T.

Douglas, R. W. (1966). *Br. J. Appl. Phys.* **17**, 435–48, see p. 443.

Ernsberger, F. M. (1968). *J. Am. Ceram. Soc.* **51** (10), 545–547.

Fayet, A. (1969). *Vetro Sili* **13** (76), 5–13.

Fraser, D. B., and LeCraw, R. C. (1964). *Rev. Sci. Instrum.* **35** (9), 1113–1115.

Gilman, J. J. (1975). *J. Appl. Phys.* **46** (4), 1625–1633.

Grosskopf, K., and Scholze, H. (1969). *Naturwissenschaften* **56**, 85.

Hashin, Z., and Shtrikman, S. (1963). *J. Mech. Phys. Solids* **11** (2), 127–140.

Hillig, W. B. (1962). *Proc. Symp. Resistance Mecanique verre, Moyens Ameliorer* pp. 295–325. Union Scientifique Continentale du Verre, Charleroi.

Huang, Y. Y., Hunt, J. L., and Stevens, J. R. (1973). *J. Appl. Phys.* **44** (8), 3589–3592.

Inamura, T. (1967). *Denki Isushin Kenkyusho Kenkyu Jitsuyoka Hokoku* **16** (5), 1003–1010; *Chem. Abstr.* **68**, 24258 (1969).

Mackenzie, J. D. (1963). *J. Am. Ceram. Soc.* **46**, 461–470.

Mackenzie, J. D. (1972). *Proc. Int. Conf. Mech. Behavior Mater., 1971* **4**, 347–359. The Society of Materials Science, Japan.

Makishima, A., and Mackenzie, J. D. (1973). *J. Non-Cryst. Solids* **12** (1), 35–45.

Mallinder, F. P., and Proctor, B. A. (1964). *Phys. Chem. Glasses* **5**, 91–103.

Manghnani, M. H. (1972). *J. Am. Ceram. Soc.* **55** (7), 360–365.

Manghnani, M. H. (1974). Pressure and Temperature Studies of Glass Properties Related to Vibrational Spectra, Final Report, Contract N00014-67-A-0387-0012, NR032-527. Hawaii Institute of Geophysics, December.

Manghnani, M. H., Murli H., and Benzing, W. M. (1969). *J. Phys. Chem. Solids* **30** (9), 2241–2245.

Meitzler, A. H., and Fitch, A. H. (1969). *J. Appl. Phys.* **40** (4), 1614–1621.

Mizouchi, N., and Cooper, A. R. (1971). *Mater. Sci. Res.* **5**, 461–476.

Morey, G. W. (1954). "The Properties of Glass," Chapter XII. Van Nostrand-Reinhold, Princeton, New Jersey.

Neely, J. E. (1969). Dissertation, Rensselaer Polytech. Inst. Univ. Microfilms, 70-2634.

Neely, J. E., and Mackenzie, J. D. (1968). *J. Mater. Sci.* **3** (6), 603–609.

Nemilov, S. V., and Gilev, I. S. (1972). *Inorg. Mater.* **8**, 294–297.

Peter, K. W. (1970). *J. Non-Cryst. Solids* **5** (2), 103–115.

Phillips, C. J. (1964). *Glass Technol.* **5** (6), 216–223.

Poch, W. (1967). *Phys. Chem. Glasses* **8**, 129–131.

Postnikov, V. S., Baleshov, Yu. S., Chernyshov, A. V., Udovenko, N. G., and Nikulin, V. kh. (1971). *Inorg. Mater.* **7**, 425–428.

Prod'homme, M. (1968). *Phys. Chem. Glasses* **9**, 101–106.

Ray, N. H. (1971). *Proc. Int. Congr. Glass, 9th* **1**, 655–663. Institut du Verre, Paris.

Robredo, J., Calvo, M. L., and Dusollier, G. (1970). *Bol. Soc. Esp. Ceram.* **9**, 9–33.

Sanditov, D. S., and Bartenev, G. M. (1973). *Russ. J. Phys. Chem.* **47** (9), 1261–1263.

Shand, E. B. (1958). "Glass Engineering Handbook," 2nd ed., p. 43. McGraw-Hill, New York.

Shaw, R. R., and Uhlmann, D. R. (1971). *J. Non-Cryst. Solids* **5** (3), 237–263.

Simmons, J. H., Macedo, P. B., Napolitano, A., and Haller, W. K. (1970). *Discuss. Faraday Soc.* No. 50 155–165.

Sinha, N. K. (1977). *J. Mater. Sci.* **12**, 557–562.

Skolowski, T. J., and Manghnani, M. H. (1969). *J. Am. Ceram. Soc.* **52** (10), 539–542.

Soga, N., and Anderson, O. L. (1967). *J. Geophys. Res.* **72**, 1733.

Soga, N., and Anderson, O. L. (1965). *Proc. Int. Congr. Glass,* 7th paper No. 37. Institute of Verre, Paris.

Soga, N., and Schreiber, E. (1968). *J. Am. Ceram. Soc.* **51** (8), 465–466.

Sosman, Robert B. (1927). *American Chemical Society Monograph* No. 37, p. 445. Chemical Catalog Co., New York.

Szigeti, B. (1950). *Proc. R. Soc. London Ser. A* **204**, 51–62.

Taylor, E. W. (1949). *Nature (London)* **163**, 323.

Westbrook, J. H. (1960). *Phys. Chem. Glasses* **1**, 32–36.

Williams, M. L., and Scott, G. E. (1970). *Glass Technol.* **11** (3), 76–79.

Yamane, M., and Sakaino, T. (1974). *Glass Technol.* **15** (5), 134–136.

Zarzycki, J. (1974). *C. R. Acad. Sci. Paris Ser. C* **278** (8), 487–490.

CHAPTER 2

Fracture Mechanics of Glass

S. W. Freiman

FRACTURE AND DEFORMATION DIVISION
NATIONAL BUREAU OF STANDARDS
WASHINGTON, D.C.

I. Introduction

Why fracture mechanics? What can fracture mechanics tell us about the properties of a material and the likelihood of failure of a structure that cannot be determined through more commonly used strength tests and statistical analysis of failure? Basically, fracture mechanics gives us a way of separating the properties of the material from the flaw distribution in the structure. Flaws in glass can be either extrinsic, e.g., due to machining or handling, or can be intrinsic, e.g., bubbles and striae. In essence, fracture

21

mechanics involves introducing a large crack into a material and determining the resistance of the material to the propagation of this crack under different environmental and stressing conditions. Knowledge of the crack propagation parameters and behavior can then be combined with information regarding flaw severity to provide a means for predicting strength and/ or lifetime of a particular structure.

It is the purpose of this chapter to provide an introduction to the principles of fracture mechanics, and then to show how the information obtained by various fracture mechanics techniques can be used to predict material behavior that is critical for glass used in structural applications. Of special importance is the phenomenon known as delayed failure, which is caused by the subcritical growth of cracks, usually in the presence of water. The use of slow crack growth data obtained by various techniques will be shown to be critical to the prediction of safe lifetimes of structures under load. In addition, it will be shown that fractographic analysis of failed specimens or parts can provide significant information regarding both the stress at fracture as well as some of the crack growth parameters.

II. Background

On an atomistic level, fracture occurs when the bonds between the atoms in a material are stretched past the breaking point. As shown in Fig. 1, the

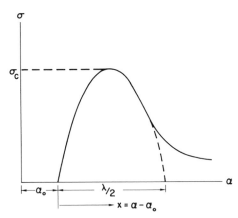

FIG. 1. Stress required to separate atomic planes; a_0 is the equilibrium separation at zero stress.

stress required to separate atom planes increases to a maximum and then decreases. Once the point of maximum stress is reached, fracture can be considered to have occurred. One can calculate the stresses required for fracture based on knowledge of the stress-displacement function for a particular material. The stress-displacement curve can be approximated by a sine curve having a wavelength λ as shown in Fig. 1, so that

$$\sigma = \sigma_c \sin (2\pi x/\lambda), \tag{1}$$

where x is the displacement from equilibrium. At small displacements, x is approximately equal to $\sin x$ so that

$$\sigma \cong \sigma_c \, 2\pi x/\lambda. \tag{2}$$

If we assume that the solid is completely elastic, then

$$\sigma = E\epsilon = Ex/a_0, \tag{3}$$

where ϵ is the strain and a_0 is the equilibrium atomic spacing, then

$$\sigma_c = \lambda E/2\pi a_0. \tag{4}$$

Since two new surfaces are created each time a bond is broken, the surface energy is given by one-half the area under the stress displacement curve:

$$\gamma_s = \frac{1}{2} \int_0^{\lambda/2} \sigma_c \sin \frac{2\pi x}{\lambda} \, dx = \frac{\lambda \sigma_c}{2\pi}, \tag{5}$$

and combining Eqs. (4) and (5), one obtains

$$\sigma_c = (E\gamma_s/a_0)^{1/2}. \tag{6}$$

If we take typical values of the parameters on the right-hand side of Eq. (6) for glass, e.g., $E = 7 \times 10^4$ MPa $a_0 = 30$ nm, $\gamma_s = 5$ J/m^2, then $\sigma_c = \approx E/7$, or 1×10^4 MPa. Strengths of this order of magnitude have been measured in pristine glass fibers by Procter *et al.* (1967). However, strengths of glasses under most conditions are 10–100 times lower than this value, because of the presence of processing-, machining-, or handling-induced flaws. These defects, which in the past have been referred to as Griffith flaws, usually occur in the form of small surface cracks. The intensification of the applied load by these flaws can be shown to lead to stresses of 10^4 MPa or greater in the vicinity of the crack tip.

Inglis (1913) showed that when a load is placed on a structure a stress concentration is produced at a crack tip. He demonstrated that the maximum tensile stress at the tip of an elliptical crack, having length $2c$ and minor axis $2h$ is given by

$$\sigma_{\max} = \sigma[1 + (2c/h)]. \tag{7}$$

When $\sigma_{\max} = \sigma_c$, crack growth will occur. This means that cracks will propagate, and failure will occur at much lower applied stresses than those theoretically required. It is to the propagation of these cracks that fracture mechanics addresses itself.

Fracture mechanics expressions used to explain crack growth behavior can be approached from either a stress concentration or an energy point of view; these two approaches are equivalent and lead to the same equations for failure.

Griffith (1921), working with glass rods, showed that the stored elastic strain energy could be equated to the surface energy formed by the propagation of a crack, as shown below:

$$4c\gamma_s = \pi\sigma^2 c^2/E \qquad (8)$$

so that

$$\sigma = (2E\gamma_s/\pi c)^{1/2}. \qquad (9)$$

In Griffith's analysis, it was the thermodynamic surface energy of the material that was considered to be important. We now know that, due to the use of energy in processes other than the formation of new surface, the energy required for rapid failure is about a factor of 10 greater than the thermodynamic surface energy estimated by extrapolation of liquid surface energy. However, the form of the expression derived by Griffith is correct and widely used.

Let us now look at fracture from a stress intensity point of view. Building on the work of Inglis and others, Sneddon (1946) showed that the stresses at a distance r from a crack tip are given by

$$\sigma_{xx} = \sigma(c/2r)^{1/2}(f(\theta)), \qquad (10a)$$
$$\sigma_{yy} = \sigma(c/2r)^{1/2}(f(\theta)), \qquad (10b)$$
$$\sigma_{xy} = \sigma(c/2r)^{1/2}(f(\theta)), \qquad (10c)$$

where θ and r are defined in Fig. 2. Irwin (1958) defined the product $\sigma c^{1/2}$

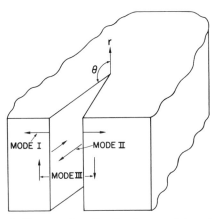

FIG. 2. Three modes of loading for a crack in a plate.

as the stress intensity factor K, which is a measure of the intensification of the applied stress due to the presence of a crack. For a crack in an infinitely wide plate, the stress intensity factor K is

$$K = \sigma(\pi c)^{1/2}. \tag{11}$$

Stress intensity factors can be defined for more-complicated geometries by various mathematical techniques. The form of the stress intensity factor also depends on the type of loading applied to the crack. As shown in Fig. 2, K_I refers to an applied tensile stress perpendicular to the crack face K_{II} to a shear force applied along the crack face in the direction of crack motion, and K_{III} to a shear force along the crack face perpendicular to its direction of motion. Typical formulations for stress intensity factor are given in Table I. For an elliptical crack, the highest stress intensity occurs

TABLE I

STRESS INTENSITY FACTORS FOR DIFFERENT CRACK GEOMETRIES

Slit crack in an infinite plate

$K_I = \sigma(\pi c)^{1/2} = (G_I E)^{1/2}$
$K_{II} = \tau(\pi c)^{1/2} = [G_I G/(1 - \nu^2)]^{1/2}$
$K_{III} = \tau(\pi c)^{1/2} = (G_I G)^{1/2}$

Imbedded crack, semiminor axis a, and semimajor axis b
$K_I = \sigma(\pi a)^{1/2}/\phi$

ϕ is an elliptical integral $= \displaystyle\int_0^{\pi/2} [1 - (b^2 - a^2)b^{-2} \sin^2\theta]^{1/2} \, d\theta$

Slit crack: $a/b = 0$, $\phi = 1.0$; penny crack: $a/b = 1$, $\phi = \pi/2$

Surface crack, semiminor axis a, and semimajor axis b
$K_I = 1.12 \, \sigma(\pi a)^{1/2}$
1.12 is the free-surface correction.

at the ends of the minor axes of the ellipse. The crack size is always taken as the smaller dimension of the crack. The other dimension is accounted for by the elliptical integral ϕ, defined in Table I. Irwin also coined the term strain energy release rate G_I, defined as the change in strain energy in a body that occurs with a change in crack length. He showed that fracture would occur when $G_I (= K^2/E)$ reaches a critical value. For an elastic crack in an infinitely wide plate,

$$G_{Ic} = \sigma_f^2 \pi c/E, \qquad \sigma_f = (EG_{Ic}/\pi c)^{1/2}. \tag{12}$$

Comparing this with Eq. (9), it is seen that, by definition, $G_I = 2\gamma$.

III. Fracture Mechanics Measurement Techniques

As noted earlier, the measurement of fracture mechanics parameters involves the introduction and subsequent propagation of a large crack in a material. In glasses this procedure is simplified because these materials can generally be treated as a homogeneous continuum. The measurement can be approached from either a stress intensity or a strain energy release rate point of view. The strain energy release rate approach is usually somewhat simpler since it does not require a detailed stress analysis. Instead, one determines the change in the stored energy in the specimen with an increase in crack length, through a measurement of the specimen compliance, which is defined as the loading point displacement divided by the load. A general relationship between G_I and compliance is obtained as follows:

Assume a crack is propagating under fixed load conditions; the load-deformation curve will resemble that in Fig. 3. It can be seen that the

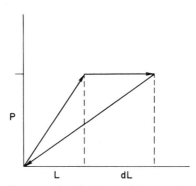

FIG. 3. Load deformation curve for a body containing a crack; fixed load conditions.

energy expended in propagating the crack a distance dL is

$$dW = 1/2PL + PdL - 1/2P(L + dL)$$
$$= 1/2PdL$$
$$= 1/2P^2 \, d(L/P), \tag{13}$$

so that

$$\frac{dW}{dL} = 1/2P^2 \left(\frac{dC}{dL}\right), \tag{14}$$

where C is the compliance of the specimen (i.e., displacement L per load P). If Eq. (14) is normalized for the specimen thickness, then

$$G_I \equiv \frac{dW}{dLt} = \frac{P^2}{2t} \left(\frac{dC}{dx}\right). \tag{15}$$

It can be shown that a generalized loading configuration also leads to Eq. (15). A general technique for obtaining G_{Ic} on any specimen geometry is to measure the compliance of a specimen at a number of different crack lengths.

A. SPECIMEN GEOMETRIES

1. Double Cantilever Beam Specimen

There are a number of loading geometries for which analytical expressions for G_I are available. These include the three configurations shown in Fig. 4 that are varieties of the double cantilever beam technique. All of

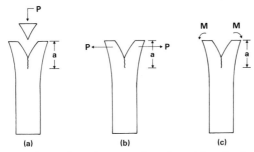

FIG. 4. Common double cantilever beam configurations: (a) wedge loaded, $G_I \propto 1/a^4$; (b) loaded in tension, $G_I \propto a^2$; (c) applied moment, G_I independent of a.

these techniques have the advantage of requiring only small amounts of material and specimens that are relatively simple to load. In addition, there are particular advantages and disadvantages to each test configuration shown in Fig. 4. The wedge loaded, constant displacement technique (Fig. 4a) is a stable configuration in that G_I decreases with increasing crack length. A number of measurements can therefore be made on the same specimen. However, in order to obtain accurate values of G_I, the displacement h and crack length a must be accurately measured. However, because K_I is difficult to determine accurately, this technique does not lend itself to crack growth measurements, although a modification of this technique developed by Kobayashi et al. (1978) suggests a way of doing this. Their technique involves tapering the cantilever arms such that G_I becomes independent of crack length. Under these conditions, a velocity–K_I plot can be obtained from the load relaxation curve, provided frictional effects can be accounted for. Configuration b is the most commonly known version of this test and has been used extensively in the study of crack growth in glass. While it is basically an unstable configuration, $G_I \propto a^2$, sufficient control can be exercised to yield crack velocity measurements. Usually holes are drilled through the glass arms and the load applied through pins in a test machine. This specimen can also be tapered so that G_I is independent of crack length, but this can significantly increase the fabrication costs. One must also correct the value of G_I to account for shear and beam rotation effects.

Configuration c provides a specimen in which G_I is independent of crack length that can be determined either visually or from a measurement of the deflection of the specimen loading arms (Freiman *et al.*, 1973). These arms are either clamped or cemented to the specimen. No corrections for shear or rotation are required when using this configuration. In general, a center groove is required for double cantilever specimens, in order to keep the crack from growing out the side of the specimen.

2. Double Torsion Technique

This test, shown schematically in Fig. 5, was first developed by Outwater and Gerry (1966) to be used under constant load conditions. Its importance

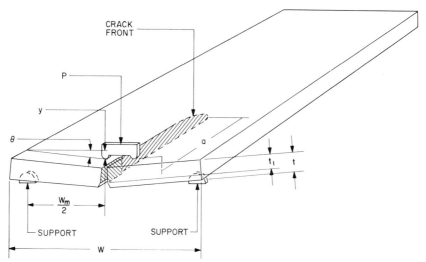

FIG. 5. Double torsion test specimen. $G_I = 3P^2 W_m^2/\{8t^3 t_1 WG[16/3 - 6.72(t/w)]\}$

was greatly magnified by the work of Williams and Evans (1973), who showed that under constant displacement or displacement rate, both G_I and crack velocity data could be obtained without the need to observe the crack directly. As demonstrated in Fig. 5, another advantage of this test is the absence of a need for grips or attachments to the specimen. This factor is especially important for elevated temperature testing. A center groove is usually placed in this specimen to guide the crack. Two disadvantages of this specimen are the large amount of material required and questions regarding the effect of the curved crack front on crack growth. A recent review of this technique by Fuller (1979) discusses some of the complications that may arise in the use of the double torsion technique.

3. Notched Bend Test

Because of the small amount of material required and the simplicity of the test procedure, the notched bend test is relatively popular. The analytic solution for crack length small relative to the specimen thickness (Fig. 6) is

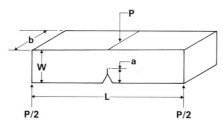

FIG. 6. Typical notched bend specimen. $K_1 = Y(3PL/2bW^2)\sqrt{a}$, where $Y = f(L/W, a/W)$.

straightforward. The major problem that arises is that unless a sharp crack can be initiated at the notch root, the measured value of K_{Ic} is strongly dependent on the notch geometry.

4. Indentation

A recently developed test involves the introduction of a hardness indentation into a test bar (Petrovic et al., 1975). By suitably choosing the load at which the indentation is produced, a crack can be produced underneath the indentation. The bar is then fractured in bending; based on the fracture stress, and the measured size and geometry of the crack, the fracture toughness can be calculated from standard fracture mechanics expressions. While the same test can essentially be performed using a bar containing flaws produced by machining or handling, the use of a hardness indentation allows for greater reproducibility of results. Evans (1979) has recently shown that K_{Ic} can be obtained by measuring the length on the tensile surface of the cracks emanating from the indent, thus eliminating the need for fracturing the specimen. A significant advantage of indentation tests, especially the latter, is that they can be used on small amounts of material and, in fact, can be used to determine the variation in K_{Ic} within a piece of material. Questions regarding possible effects of plastic deformation and residual stresses at the indentation are currently being resolved.

B. FRACTOGRAPHIC ANALYSIS

In order to determine how well fracture mechanics explains the failure of glass, it is necessary to have some means of assigning a value to the flaw severity in a specimen. The most direct method is to measure the size and geometry of the flaw on the fracture surface. As shown in Table I, for a surface flaw

$$K_{Ic} = 1.12 \, \sigma_c \sqrt{\pi a}/\phi. \tag{16}$$

With few exceptions, surface flaws produced by machining or handling are the predominant source of failure in glass, as illustrated in Fig. 7.

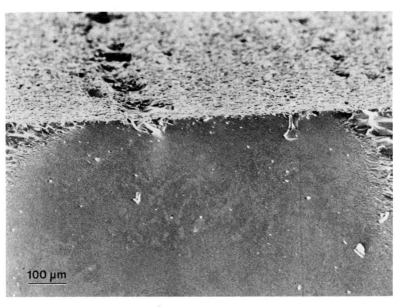

FIG. 7. Fracture surface of glass flexure bar ground parallel to the tensile axis. Failure originated from the crack introduced below the deeper than average grinding groove.

The flaw size and geometry required for Eq. (16) are the critical dimensions, i.e., those at which the flaw grows catastrophically. Unfortunately the values usually measured are taken from the initial flaw since no flaw boundary is produced on a fracture surface that corresponds to a critical flaw size. The difference between the initial and critical flaw sizes (shown schematically in Fig. 8) is determined by the amount of subcritical crack growth and is a function of material and environment. For tests performed under relatively rapid loading conditions, values of K_{Ic} calculated from the initial flaw sizes agree quite well with those measured by fracture mechanics techniques (Mecholsky *et al.*, 1974).

In addition to the flaw itself, there are other fracture surface features that are quantitatively related to the stress at failure and the crack propagation resistance of the material. As shown schematically in Fig. 8 and a typical glass fracture surface in Fig. 9, several regions can be seen on the fracture surface. Immediately surrounding the flaw is a smooth, shiny region that

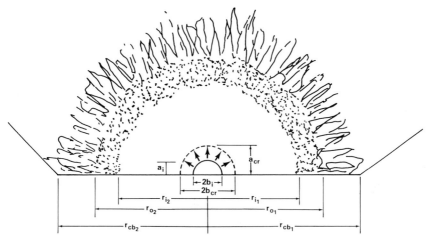

FIG. 8. Schematic of features observed on a glass fracture surface: r_i, mist boundary; r_0, hackle boundary; r_{0b}, macroscopic crack branching; a_i, initial flaw size; a_{cr}, size at which flaw extends catastrophically (no markings are left on the surface at this point). $(a_{cr}b_{cr})^{1/2}/r = \text{const}$; $\rho r^{1/2} = \text{const}$.

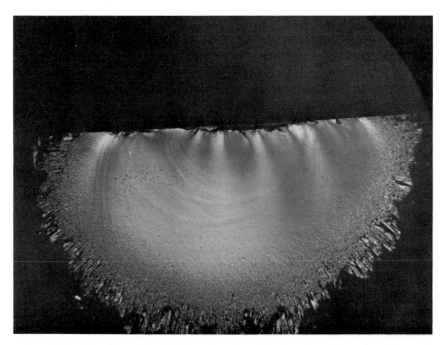

FIG. 9. Typical mist and hackle formed on glass; flaw is elongated crack at the tensile surface.

historically has been referred to as a fracture mirror (Shand, 1954). This mirrorlike region is bounded by a region containing small radial ridges known as mist, which is bounded in turn by a region containing even larger ridges known as hackle. Finally, macroscopic cracks branching occurs, i.e., the formation of more than one primary crack. It was experimentally shown some time ago (Shand, 1959) that the distance of the demarcation from the flaw is inversely related to the fracture stress, i.e.,

$$\sigma = A_j r_j^{-1/2}, \qquad (17)$$

where r_j is the distance to a particular boundary and A_j is the corresponding "mirror constant."

Despite the fact that the various boundaries do not show up as sharp lines in the microscope, the values of mirror constants for the same glass obtained over a number of years in different laboratories are quite similar (Table II). The demarcations correspond to a certain critical density of frac-

TABLE II

FRACTURE MIRROR CONSTANTS FOR SODA-LIME GLASS[a]

Mirror-mist boundary	Mist-hackle boundary
1.7 (Abdel-Latif *et al.*, 1977)	2.0 (Kirchner and Gruver, 1974)
1.7 (Bansal and Duckworth, 1977)	2.1 (Mecholsky *et al.*, 1976)
1.9 (Mecholsky *et al.*, 1976)	
1.9 (Kerper and Scuderi, 1965)	
1.9 (Johnson and Holloway, 1966)	

[a]A is expressed in MPa/m$^{1/2}$.

ture features and are best observed in the optical microscope using reflected light at a magnification such that the mirror radius $\simeq 2$ cm in the microscope. One must also be aware of edge effects both at the tensile surface and the specimen edges.

Examples of the stress–mirror size relationship for different glasses are illustrated in Fig. 10. The curves are best-fit straight lines of slope -0.5. By plotting a line having this slope, one assumes that Eq. (17) is valid, which seems to be true based on available data. Where residual stresses are present, as in the work of Kerper and Scuderi (1965), the logarithmic plot of σ versus r is no longer linear. However, if one plots σ versus $r^{1/2}$ as shown in Fig. 11, a straight line having a stress intercept of 70 MPa is obtained. This intercept value is a reasonable estimate of the residual stress in this glass (Mecholsky *et al.*, 1978).

Equations (16) and (17) can be combined to relate quantitatively the fracture mirror boundaries to the fracture toughness of the material:

$$A = \phi \, K_{Ic}/(1.2 \, \pi \, c/r)^{1/2} \qquad (18)$$

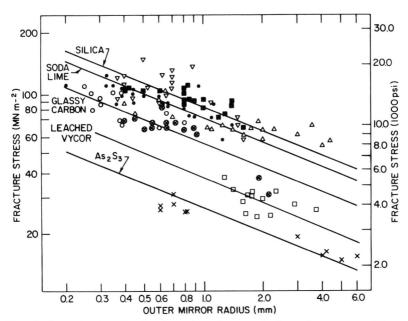

FIG. 10. Strength-mirror size data for various glasses x, As₂S₃; ○, glassy carbon; ■, borosilicate; •, soda lime; Δ, SiO₂ glass; ∇, aluminosilicate; •, lead silicate; □, leached Vycor. (Mecholsky *et al.*, 1974.)

where $c = (a\,b)^{1/2}$. If A is plotted versus K_{Ic}, data for glasses (as well as polycrystalline ceramics) fall in a band about a hackle to flaw size ratio of 13 (Mecholsky *et al.*, 1976).

It is generally thought that mist and hackle represent the initiation and limited propagation of secondary cracks, which are formed because of the excess kinetic energy available as the primary cracks approach a terminal velocity of ~0.6, the shear wave velocity in the glass. Three different criteria have been used to try to explain the formation of mist and hackle. These would predict that secondary crack formation occurs at (1) a characteristic or critical velocity (Yoffee, 1951), (2) a particular stress intensity or strain intensity (Clarke and Irwin, 1966; Mecholsky *et al.*, 1976; Kirchner, 1976), or (3) a strain energy release rate greater than that required for planar cracks (Mott, 1948; Roberts and Wells, 1954; Abdel Latif *et al.*, 1977).

Mecholsky and Freiman (1979) have compared the values of mirror constant A with those predicted by several of the above theories. As shown in Table III, none of the predictions is in particularly good agreement with experiment. Part of the reason for this lack of agreement is the lack of sensitivity of crack velocity to fairly large changes in crack length.

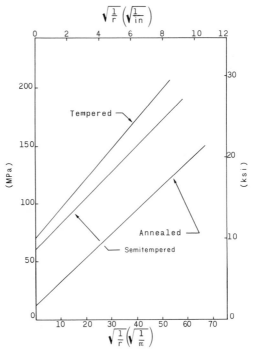

$$\sqrt{\tfrac{1}{r}}\left(\sqrt{\tfrac{1}{in}}\right)$$

FIG. 11. An aluminosilicate glass replotted as strength vs. (1/mirror radius)$^{1/2}$. Positions of lines on the ordinate represent differences in the residual stresses in the glass (data from Kerper and Scuderi, 1965).

IV. Critical Fracture Toughness of Glass

The first parameter that comes to mind when one considers approaching the fracture of glass from a fracture mechanics point of view is the critical fracture toughness of the material, K_{Ic}. As shown in a previous section K_{Ic} is directly related to the critical strain energy release rate G_{Ic} and the critical fracture energy γ_c ($\equiv G_{Ic}/2$). A point to be made is that K_{Ic} is not necessarily a material's property in the exact sense of the term but is dependent on parameters such as the test environment and the speed at which the test is conducted. K_{Ic} is really the point on a crack velocity-stress intensity factor (V–K_I) diagram at which the specimen compliance decreases due to crack growth such that the measured load drops off, e.g., about 10^{-1} m/sec under normal test conditions. Since, as will be shown later, the V–K_I curve is quite steep at this point, variations in K_{Ic} are usually small.

Critical fracture energy measurements on Pyrex and float glass were first made by Linger and Holloway (1968) (Table IV). Wiederhorn (1969) expanded work on the critical fracture energy of glass to include other glass

TABLE III

COMPARISON OF CRITERIA FOR CRACK BRANCHING IN SODA-LIME GLASS

		A/K_{lc}		
Author(s)	Formulation[a]		Boundary	Measured
Johnson and Holloway (1966)	$\left[\dfrac{4}{2\pi - (K\rho V^2/E)}\right]^{1/2}$		(mirror-mist) (mist-hackle) 1.2 (branching)	2.5 2.7
Congelton and Petch (1967)	$(\sqrt{2}/\pi)$		0.8 (branching)	1.4
Abdel-Latif et al. (1977)[b]	$\left[\dfrac{2}{3[1 - (K\rho V^2/E)]}\right]^{1/2}$		1.8 (mirror-mist)	2.3
Bansal (1978)	$\left[\dfrac{2\pi}{8 - (3K\rho V^2/2E)}\right]^{1/2}$		1.6 (mirror-mist)	2.3
Mecholsky et al. (1978)	$\left[\dfrac{1.25}{2(c/r)}\right]^{1/2}$		2.8 (mirror-mist) 3.2 (mist-hackle)	2.7 3.1

[a]$V = 1500$ m/sec; $\rho = 2.5$ gm/cm^2; $K = 44$.
[b]Tensile case; cannot use $K = 44$ or $V = 1500$ m/sec because the term in brackets becomes negative, so $K = 22$ and $V = 1000$ m/sec was used as suggested in the reference.

TABLE IV

CRITICAL FRACTURE ENERGY OF GLASS

Glass	Environment	$\gamma_c(j/m^2)$	Reference
Pyrex	Air, 20°C, 20% RH	4.7	Linger and Holloway (1968)
	Air, 22°C, 40% RH	4.0	Mecholsky et al. (1974)
	N$_2$(gas), 27°C, <0.1% RH	4.5–4.8	Wiederhorn (1969)
	N$_2$(l), 77 K	4.7	Wiederhorn (1969)
	Water, 20°C	2.5	Linger and Holloway (1968)
Soda-lime (or float glass)	Air, 20°C, 20% RH	3.9[a]	Linger and Holloway (1968)
	N$_2$(l), 77 K	4.1	Linger and Holloway (1968)
	N$_2$(l), 77 K	4.5–4.6	Wiederhorn (1969)
	N$_2$(gas), 27°C, <1% RH	3.8–3.9	Wiederhorn (1969)
	Air, 22°C, 40% RH	3.5	Mecholsky et al. (1974)
	Vacuum, 10^{-4} Torr	5.0	Linger and Holloway (1968)
Fused silica	N$_2$(gas), 27°C, <1% RH	4.3–4.4	Wiederhorn (1969)
	N$_2$(l), 77 K	4.6	Wiederhorn (1969)
	Air, 22°C, 40% RH	3.7	Mecholsky et al. (1974)
Aluminosilicate	N$_2$(gas), 27°C, <1% RH	4.6–4.7	Wiederhorn (1969)
	N$_2$(l), 77 K	5.2	Wiederhorn (1969)
	Air, 22°C, 40% RH	3.7	Mecholsky et al. (1974)

[a]Taken as the γ(2-min) value in the reference.

compositions (Table IV). Wiederhorn showed that temperature had little effect on γ_c, and demonstrated a direct correlation between elastic modulus and critical fracture energy for the group of glasses studied. Wiederhorn and Johnson (1971) showed that hydrostatic pressure had no measurable effect on critical fracture energy.

V. Application of Fracture Mechanics to Glass Fracture

Since another chapter in this volume is specifically concerned with the strength of glass, this chapter will concentrate only on those aspects of strength that can be addressed by fracture mechanics. The discussion will include the following topics: the relationship of flaw severity to fracture, the measurement of fracture mechanics parameters involved in delayed failure, the use of fracture mechanics for prediction of impact and erosion resistance, and methods of lifetime prediction for glass under stress.

A. STRENGTH–FLAW SIZE RELATIONS

The proof of the pudding as far as the strength of glass is concerned is how well can the fracture stress be predicted if we know the fracture toughness K_{Ic} and flaw size a, and geometry. Mecholsky *et al.* (1977) broke glass bars either parallel or perpendicular to the direction of grinding. It was observed that the flaws causing failure in the set of specimens tested parallel to the grinding direction were smaller and less elongated than those in the set tested perpendicular to the grinding direction (Fig. 12). Referring to Eq. (16), it was shown that for flaws in the plane of fracture, the ratio of $\phi/\sqrt{a}$ for the two sets of flaws was 0.69. The actual ratio of strengths for the two directions was 0.70. This agreement indicates the effectiveness of fracture mechanics formulations in explaining strength variations in a material. However, many of the flaws had very rough surfaces, so that parts of them were out of the plane of fracture, e.g., Fig. 13. In these cases fracture mechanics predictions underestimated the fracture strength. However, even in these cases the fracture stresses calculated from the fracture mirror size measurements were accurate.

Following up on the above study, Freiman *et al.* (1979) used a technique introduced by Petrovic and Mendiratta (1976, 1977) to examine the effect of flaw orientation on fracture in glass. Cracks were introduced into glass bars using a Knoop indentor. By varying the orientation of the indentor relative to the tensile axis of the specimens, cracks were formed with crack planes ranging from 90° to 20° to the tensile axis. These cracks propagated under combined mode I–mode II loading (Fig. 14). Of the various models for mixed mode fracture, a noncoplanar, maximum strain energy release rate model (Hussain *et al.*, 1974) fit the data over the widest range of crack

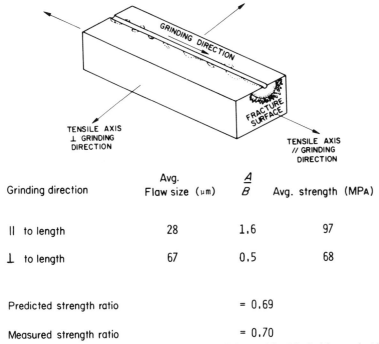

Grinding direction	Avg. Flaw size (μm)	$\frac{A}{B}$	Avg. strength (MPa)
‖ to length	28	1.6	97
⊥ to length	67	0.5	68

Predicted strength ratio   = 0.69

Measured strength ratio   = 0.70

FIG. 12. Effect of grinding direction on the strength of glass. (After Mecholsky *et al.*, 1977.)

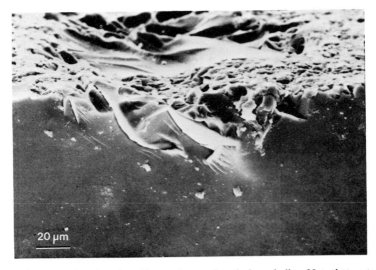

FIG. 13. Irregular flaw introduced into a glass surface during grinding. Note that parts of the flaw are not perpendicular to the tensile stress.

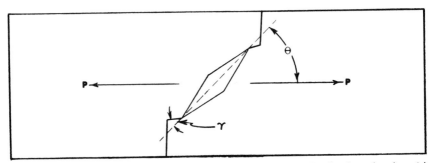

FIG. 14. Knoop hardness indentation oriented for mixed mode loading in a glass bar. θ is the angle between the original crack and the tensile axis; γ is the angle between the original crack plane and the direction of initial propagation.

angles (Fig. 15). This model predicts that a crack under a mixed mode load will initially propagate out of its plane at an angle γ that maximizes the strain energy release rate. This type of behavior was observed for the indented glass bars. In addition, it was found that the K_c calculated from the fracture mirror measurements agreed with that measured by a double cantilever beam technique over the whole range of crack angles investigated (Fig. 15). Many of the fracture mirrors were unsymmetric about the

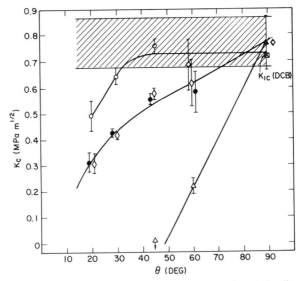

FIG. 15. Combinations of K_I and K_{II} based on different mixed mode loading models. Note that fracture mirror data yield the same value of K_c as a double cantilever beam test over the entire angular range tested. Δ, maximum stress; $\bullet$, coplanar; $\circ$, noncoplanar; $\diamond$, strain energy density; $\square$, average mirror constants. (After Freiman et al., 1979.)

failure origin, presumably because one portion of the crack began to grow before the stress intensity factor at the other portions of the crack reached a critical value. In these cases the mirror constant associated with the largest dimension of a mirror boundary was found to be correct. The largest mirror boundary is associated with crack growth that begins at the lowest stress in the bar.

Mecholsky *et al.* (1979) showed that if fracture mechanics parameters are known for a material (e.g., soda-lime glass) failing after a time under load, an estimate of this time can be obtained from the ratio of the mirror size to the initial flaw size, as shown in Fig. 16. Meanwhile, the ratio of the

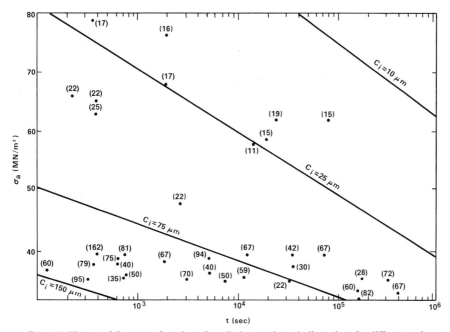

FIG. 16. Time-to-failure as a function of applied stress in soda-lime glass for different values of initial flaw size. Data points represent times-to-failure predicted from mirror size and initial flaw size measured on that specimen (given in parentheses). (After Mecholsky *et al.*, 1979.)

mirror size to the critical flaw size calculated for this glass was found to be a constant.

Fracture mechanics can also be used to predict the fracture stress of high-strength optical fibers. At fracture stresses of ~1200 MPa or less, the failure origin can generally be found even though flaw sizes are much less than 1 μm. It has been shown that the strength-mirror size curve obtained for high-strength fibers is a direct extension of the curve obtained for bulk-fused silica. This means that fracture stresses can be determined from the fracture mirror measurements made on high-strength glass fibers, even when the flaws themselves are not distinct.

B. SUBCRITICAL CRACK GROWTH

The application of fracture mechanics to explain the rapid failure of glass in a strength test has been discussed. Of perhaps even more importance is growth of cracks at subcritical velocities leading to the time-dependent failure of glass under load. In particular, fracture mechanics techniques have been used (1) to understand the fracture mechanism and (2) to develop lifetime prediction techniques. In order to show how fracture mechanics can be used to analyze this behavior, it is first necessary to review the work on subcritical crack growth and to develop the analytical expressions that will later be applied to glass failure.

In a now-classic paper, Wiederhorn (1967) measured crack velocities as a function of stress intensity in double cantilever beam specimens held in N_2 gas containing varying quantities of H_2O. He showed that the overall crack growth curve could be divided into three regions (Fig. 17). Based on Eq. (19), which he derived from chemical rate theory, Wiederhorn demonstrated that in region I, crack velocities should be governed by the rate of stress corrosion at the crack tip, i.e.,

$$\text{crack velocity} = V = (Ax_0 n \exp bK_1)/n, \tag{19}$$

where x_0 is the partial pressure, i.e., RH of water in the environment, n the order of the chemical reaction, and A and b are constants. He showed that a plot of log V versus log RH was linear, with slope n, and that n varied from $\frac{1}{2}$ to 1 with increasing RH, indicating a change in the crack tip reaction process.

Wiederhorn showed that the crack velocity in region II is essentially independent of stress intensity and should be governed by the rate of diffusion of the corroding species to the crack tip:

$$V = CD_{H_2O}X_0/n, \tag{20}$$

where D_{H_2O} is the diffusivity of H_2O in the environment, X_0 a boundary layer thickness, and C a constant. A plot of log V versus log RH in this region gave a straight line of slope 1, as predicted by the model. Recent work by Quackenbush and Frechette (1978) and Chandan *et al.* (1978) has suggested that the occurrence of a region II is dependent on crack size. That is, if the crack size is reduced sufficiently, water can always diffuse to the crack tip at a rate sufficient to provide a stress corrosion environment.

Wiederhorn's results indicated that once region III behavior is achieved, there is no longer a dependence of crack velocity on relative humidity.

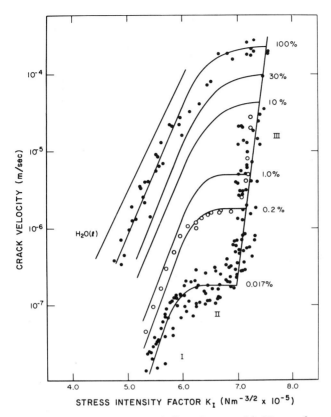

FIG. 17. Crack velocity–K_I curves for soda-lime glass tested in N_2 gas of varying relative humidity. (After Wiederhorn, 1967).

However, no crack growth governing mechanism was suggested for this region.

Freiman (1974) showed that the above model applied equally well to water dissolved in other liquids such as straight chain alcohols ranging from hexanol (C-6) to dodecanol (C-12) (Fig. 18). The fact that all of the alcohols fell on the same region I and region II curves, regardless of the variation in chain length and solubility of water in each alcohol, emphasized the point that it is the partial pressure (RH) rather than the absolute quantity of water that controls crack growth. In region III, crack velocity was independent of RH, but interestingly enough was dependent on alcohol chain length (Fig. 19).

The mechanisms governing crack growth rates in region III are clearly not understood. Reported activation energies for crack growth in this

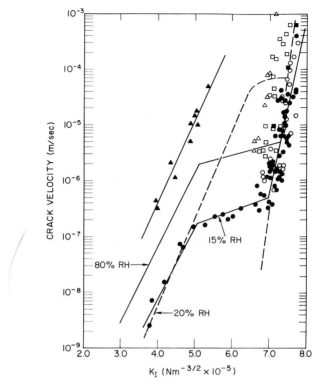

FIG. 18. Crack velocity–K_I data for soda-lime glass tested in alcohols of different chain length. Data points for all alcohols scattered about the same region I and region II curves. ▲, H_2O; Δ, hexanol; □, heptanol; •, octanol; ○, decanol; ■, dodecanol; —Wiederhorn data for moist N_2. (After Freiman, 1974.)

region range from 17 (Freiman, 1975) to 176 kcal/mole (Wiederhorn et al., 1974b), the values depending to a great degree on the form to which the data were fit. None of the theories explaining crack growth in this region appears to be entirely satisfactory. This point will be discussed further in a later section.

Wiederhorn and Bolz (1970) showed that the $V–K_I$ in region I was a function of glass composition as shown in Fig. 20. It can be seen that at low velocities there appears to be a fourth region of behavior for soda-lime and borosilicate glasses. Whether this region IA represents a "stress corrosion limit" is not clear, and this behavior is not observed for all glasses. A similar observation was made for borosilicate glass in a HCl solution of pH 1.7 (Wiederhorn and Johnson, 1973) but not at other pH values. Pyrex and fused silica tested in air of 50% RH showed no evidence of a stress corrosion limit even at velocities as low as 10^{-11} m/sec.

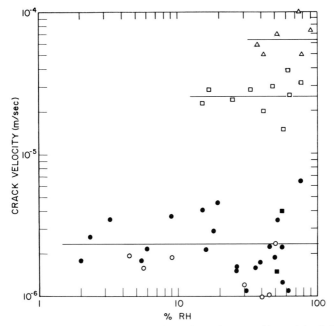

FIG. 19. Region III crack velocity as a function of RH glass tested in straight chain alcohols. Note independence of velocity on RH but dependence on alcohol chain length. △, hexanol; □, heptanol; ●, octanol; ○, decanol; ■, dodecanol. (After Freiman, 1974.)

The data in Fig. 20 suggest a trend to smaller slopes and higher crack velocities as modifying ions are added to SiO_2. However, there is no overall theory that can relate glass composition to subcritical crack growth. Data obtained by Freiman (1974) on a $BaO–SiO_2$ glass suggest that stress corrosion may not, in fact, be controlled by modifier ion concentrations in a glass (Fig. 21). The $3BaO \cdot 5SiO_2$ glass contains only a few parts per million Na_2O, yet gives rise to crack velocities at a given K_I, which are an order-of-magnitude greater than those in any of the soda-lime glass compositions. Other structural factors can also be important. For instance, Adbel-Latif *et al.* (1976) showed that phase separation in a soda-lime glass caused an increase in crack growth rates.

If crack growth is controlled by the reaction of the glass with H_2O, then one would expect that the hydrogen and hydroxyl ion concentration (pH) in the water to be an important factor. Wiederhorn and Johnson (1973) showed that the pH of the solution did affect crack growth but in a somewhat complex way. They found that the slopes of the $V–K_I$ curves for fused silica, borosilicate, and soda-lime glass depended on pH. In fused silica for instance, the slope increased by about a factor of 2 as the pH was

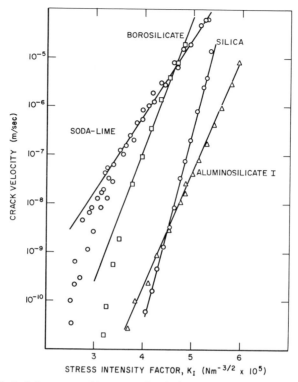

Fig. 20. Effect of glass composition on crack velocity–K_I behavior of various glasses tested in H_2O at room temperature. (After Wiederhorn and Bolz, 1970.)

varied from 14.8 (6N NaOH) to —0.8 (6N HCl). In the other two glasses, behavior was more complex, the slopes varying similar to fused silica at low velocities, but approaching a common curve at velocities greater than 10^{-5} to 10^{-4} m/sec. as illustrated in Fig. 22. As seen in Fig. 22B, dilute buffered solutions lead to a common curve for soda-lime glass over a wide range of pH. The above behavior is explained by the fact that mobile ions must be transported to the crack tip in order to affect the crack growth process. As solution moves to the crack tip, the concentration of mobile ions is modified by exchange with ions in the glass. This exchange process is velocity dependent. At low velocities, diffusion is able to maintain the concentration of the crack tip solution so that it is similar to the bulk electrolyte. At high velocities, the crack tip solution is controlled by the glass composition, as indicated by the common curve for soda-lime and borosilicate glass in this regime. In fused silica, the bulk environment always controls growth, presumably because of the relative inertness of this material. The above phenomena also explain the results of Mould (1961) and Langi-

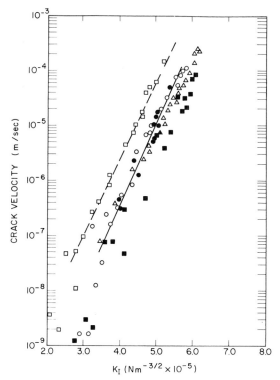

FIG. 21. Comparison of velocity–K_I behavior of soda-lime glasses (●, 1; ○, 2; Δ, 3; ■, 4) and a low alkali-containing $3BaO \cdot 5SiO_2$ glass (□, 5). (After Freiman, 1974.)

tan and Lawn (1970). Mould found that the strength of soda-lime glass was relatively insensitive to buffer solutions ranging in pH from ~1 to 13. He found that the strength increased for pH values greater than 13 and decreased for pH less than 1. Wiederhorn and Johnson's values of strength calculated from the crack velocity data agreed well with those of Mould.

In an investigation initiated to help choose a material for the windows in the manned space laboratory, Wiederhorn et al. (1974b) conducted crack growth tests in vacuum of less than 10^{-4} Torr on six glass compositions. They found that for four of these glasses, subcritical crack growth could be measured (Fig. 23). In contrast, only abrupt fracture occurred in borosilicate glass and fused silica at all temperatures investigated (up to 775°C). These latter two glasses exhibit what is termed anomolous properties, namely, a low thermal expansion coefficient, a positive temperature dependence of bulk modulus, and a negative pressure dependence of bulk modulus. An interesting point in regard to the above results is that Ernsberger (1969), in studying the fracture of glasses from internal bubbles, observed

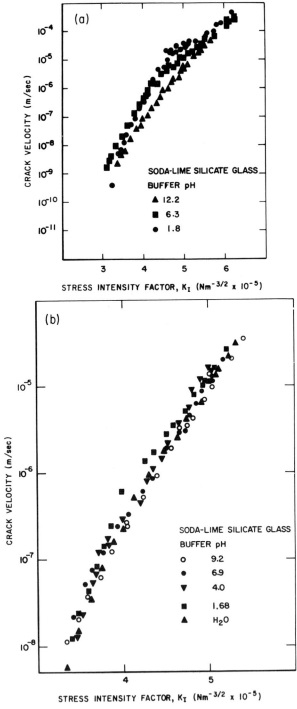

Fig. 22. Soda-lime glass tested in (a) concentrated buffer solutions, (b) dilute buffer solutions, and (c) acid, bases, and water. (After Wiederhorn and Johnson, 1973.)

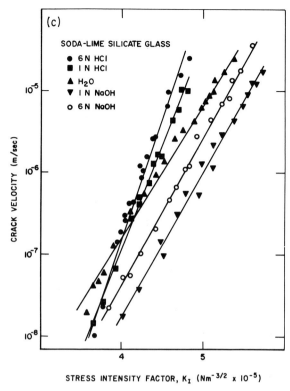

FIG. 22. *Continued*

that borosilicate glass and fused silica failed due to densification, while soda-lime glass cracked in shear. Widerhorn *et al.* (1974b) concluded that alkali ion diffusion or viscous flow models would not explain the fracture data, the former because of the high-activation energies measured for crack growth (60–175 kcal/mole), and the latter because of the large relaxation times for glass in the temperature range studied. They suggested that a lattice trapping model proposed by Thomson *et al.* (1971) (which will be discussed later) may explain the data. Tyson *et al.* (1976) attempted to show more quantitatively that the above data fit a lattice trapping model.

Wilkins and Dutton (1976) demonstrated the existence of a static fatigue limit by measuring the room temperature strength of soda-lime glass rods after specimens were held under different levels of stress at 400°C for 24 hr. Those specimens held under higher stress were reduced in strength, while those held under lower stresses were strengthened. In addition, for any given strength distribution, the upper portion was moved toward higher strengths while the lower was moved toward reduced strength. The stress

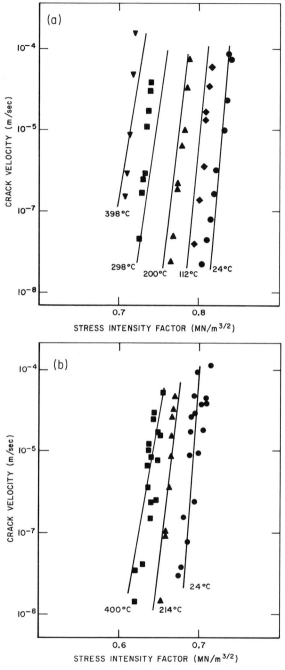

Fig. 23. Crack propagation in glass tested in a vacuum of 10^{-4} Torr as a function of temperature (a) soda-lime silica, (b) aluminosilicate, (c) borosilicate crown, and (d) 61% lead glass.

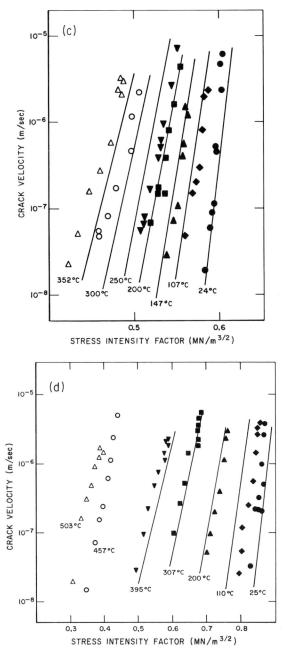

FIG. 23. *Continued*

level at which no strengthening or weakening occurred was taken as a measure of the static fatigue limit at 400°C.

C. Mechanisms of Subcritical Crack Growth

A large number of theories have been proposed over the years to explain subcritical crack growth and delayed failure in glasses. These are discussed in some detail in a review by Wiederhorn (1978). In this section a few of these theories will be discussed critically, and an attempt will be made to show how they fit into the framework of available data. In many of these models crack growth data or strength data were fit to a theoretical expression and the quality of the fit was taken as proof of the theory. Because of the large variation in crack velocity with small changes in K_I, it can be easily seen that there are many theoretical expressions that can be "fit" to the data with equal accuracy. As shown in Fig. 24 for instance, Wiederhorn

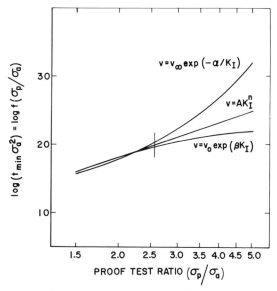

FIG. 24. Crack propagation data expressed in three different forms over a range of proof stresses corresponding to a range of crack velocities. The vertical line at σ_p/α_a of 2.5 corresponds to the lowest measured crack velocity. Extrapolation beyond this point leads to deviation of the different curve forms leading to differing lifetime predictions. ●, $6N$ HCl; $1N$ HCl; ▲, ■, H_2O; ▼, $1N$ NaOH; ○, $6N$ NaOH. (After Wiederhorn, 1977.)

(1978) demonstrated that it is almost impossible to distinguish between the fits of three diverse expressions for crack growth over the normal data range. In order to be a successful model for crack growth, a theory must also explain the temperature, environment and compositional dependence of crack growth data.

The theory that seems to have demonstrated the best capability for modeling stress-corrosion-governed crack propagation (region I) is that of Hillig and Charles (1965). They approached the problem from the point of view of chemical reaction rate theory. Wiederhorn (1972) derived an expression for crack velocity controlled by chemical reactions at the crack tip. The form of this expression is nearly identical to that presented by Hillig and Charles, namely,

$$V = V_0 \exp \left[(-\Delta E^{\ddagger} + (\sigma \Delta V^{\ddagger}/3) - V_m \gamma/\rho)/RT \right], \qquad (21)$$

where V_0 depends on chemical activity at the flaw tip, $\Delta E^{\ddagger}$ the activation energy of the chemical reaction in the absence of stress, σ the stress at the glass–liquid interface, and $\Delta V^{\ddagger}$ the activation volume. The last term on the right-hand side accounts for the change in surface curvature, where V_m is the molar volume of the glass, γ the surface tension at the glass–environment interface, and ρ the radius of curvature of the crack tip. Hillig and Charles suggested that at stresses greater than some threshold value, the crack tip radius decreases due to stress corrosion, and the crack will grow. Below this threshold stress the radius increases, leading to blunting and strengthening. The threshold value of stress can be taken as a stress corrosion limit. Equation (21) can be expressed in fracture mechanics form as

$$V = V_0 \exp \left[(-E^* + 2V^{\ddagger}K_1/\sqrt{\pi\rho})/RT \right], \qquad (22)$$

where E^* is now $E^{\ddagger} + V_m\gamma/\rho$. This expression suggests that plots of log V versus K_1 should be displaced to higher V and should decrease in slope with increasing temperature. A value for E^* can be obtained by extrapolating the V–K_1 plots to $K_1 = 0$, and plotting log V (at $K_1 = 0$) versus $1/T$. Extrapolation of V–K_1 curves over such long distances, however, can lead to substantial errors. A better approach is that taken by Wiederhorn and Bolz (1970), who fit a series of V–K_1 curves for glass in H_2O to Eq. (22) over a temperature range of 2°–90°C. They obtained the plots shown in Fig. 25 and the fitting parameters given in Table V. The fact that the individual

TABLE V

STRESS CORROSION DATA FOR GLASSES TESTED IN H_2O[a]

Glass	E^* (kcal/mole)	b (mks units)	ln $V_0^{\ddagger}$
Silica	33.1 ± 1.0	0.216 ± 0.006	−1.32 ± 0.6
Aluminosilicate I	29.0 ± 0.7	0.138 ± 0.003	5.5 ± 0.4
Aluminosilicate II	30.1 ± 0.6	0.164 ± 0.003	7.9 ± 0.3
Borosilicate	30.8 ± 0.8	0.200 ± 0.005	3.5 ± 0.5
Lead alkali	25.2 ± 1.2	0.144 ± 0.006	6.7 ± 0.6
Soda-lime silicate	26.0 ± 1.1	0.110 ± 0.004	10.3 ± 0.5

[a] After Wiederhorn and Bolz (1970).

curves in Fig. 25 are very nearly parallel suggests that the stress dependence to the activated process could be decoupled from the temperature dependence yielding an equation of the form

$$V = V_0 \exp\left(-E^*/RT\right) \exp bK_1. \tag{23}$$

Although data obtained by Wiederhorn *et al.* (1974b) for glasses propagating in vacuum did show a greater stress-dependent activation energy

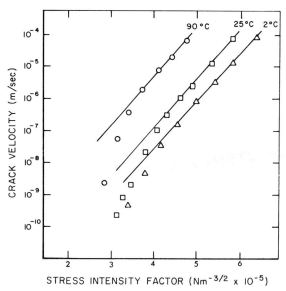

FIG. 25. Effect of temperature on the crack velocity behavior of soda-lime glass in H_2O. Note that fit was made to data above the point at which velocity decrease occurred. (After Wiederhorn and Bolz, 1970.)

for crack growth (Fig. 23), the data scatter is such that it is difficult to rule out the possibility of a fit to Eq. (23). In fact, if the soda-lime data shown in Fig. 23 is fitted to Eq. (23), an activation energy, $E^* = 12$ kcal/mole is obtained. This value for E^* was also obtained by Gerberich and Stout (1976) using a slightly different approach. Similar conclusions were also reached with regard to region III crack propagation data obtained at various temperatures in alcohols of different chain lengths (Freiman, 1975). Because the data could be fit to a number of different equations, Freiman concluded that a decision about a stress corrosion mechanism could not be arrived at solely on the basis of activation energy data. It should be emphasized that the experiments of Wiederhorn and Bolz were conducted under conditions where stress corrosion governed crack growth, (region I) whereas the latter two studies were carried out under conditions where crack growth is insensitive to water in the environment.

Gerberich and Stout (1976) took the approach that subcritical crack growth in glasses may be governed by the diffusion of an "embrittling" species to the crack tip, analogous to hydrogen embrittlement in metals. They derived the following equation for crack velocity:

$$V = [2(1 + \nu) D_A \overline{V} K_1 C_0 2 / [3 r^{3/2} RT(C_{cr} - C_0)] \tag{24}$$

where ν is Poisson's ratio, D_A the apparent diffusivity of the embrittling species, $\overline{V}$ the partial molar volume, C_0 the initial concentration of the species, C_{cr} the critical concentration, and r the distance from the crack tip. They used Eq. (24), which is similar to Eq. (22), to fit the temperature dependence of the data of Wiederhorn *et al.* (1974b) and Freiman (1974), and attempted to show that crack growth could be controlled by the diffusion of the Na^+ to the crack tip or by interdiffusion of Na^+ and H^+. However, as the authors themselves point out, Eq. (24) cannot be entirely correct, because we know the crack velocity is not linearly dependent on K_1. They felt that if concentration gradient terms are included, then a correct relationship between V and K_1 would be obtained.

Cox (1969) proposed a model suggesting that the formation of flaws in a glass can occur due to the motion of Na^+ ions under stress. While this model addresses flaw generation rather than growth, an interesting aspect of Cox's model is that the concentration of Na^+ ions is relatively unimportant, so that the behavior of soda-lime glass and fused silica containing ~ 1 ppm Na^+ could both be explained. However, Ritter and Manthuruthil (1973) concluded that an equation based on Cox's model did not fit delayed failure data.

Hasselman (1970) and Stevens and Dutton (1971), have also proposed diffusion-controlled crack growth models. Their theories are basically based on the concept that crack growth rates are controlled by the diffusion of vacancies to the crack tip. Their models predict that activation energies for crack growth would match those for diffusion of a particular species in glass. As was pointed out earlier, however, values of activation energy calculated from a set of temperature-dependent crack velocity data depend heavily on the form of the expression used to fit the data. Therefore, it is very nearly impossible to choose between different models for crack growth based on predictions of activation energies.

A recent model for crack growth based on a multibarrier kinetic theory of stress corrosion has been developed by Brown (1978). This model suggests that crack growth consists of a competitive reaction sequence, each containing a number of steps. The crack growth equation resulting from such an analysis is quite complex, i.e.,

$$V = \Omega_0 e \Omega_1 K_I + \frac{\Omega_2 e \Omega_3 K_I [1 - e^{-L(K_I - K_I*)}]}{1 + \Omega_4 e^{\Omega_3 K_I}}, \tag{25}$$

where Ω_0, Ω_1, Ω_2, Ω_3, Ω_4, L, and K_I^* are constants. Brown showed that Eq. (25) fit a large variety of crack velocity data in glasses, ceramics, and metals over the three regions of crack propagation of the type shown in Fig. 17. In order to determine whether this approach is fundamental, however, it will be necessary to relate the constants in Eq. (25) to measured values of materials' parameters.

Weidman and Holloway (1974) derived a model for crack growth behavior based upon the concept that growth of a plastic zone controlled the rate of crack propagation. They argued that dependence of glass hardness on loading in various environments (Gunsakera and Holloway, 1973) indicated the presence of a plastic zone. They derived the following expression for crack velocity:

$$V = (K_I \, AB)(2R_c/\pi)^{1/2} \exp\left[-(2\pi R_c)^{1/2}/K_I A)\right], \tag{26}$$

where A and B are constants and R_c is the critical size of the plastic zone. Zone sizes of 3–5 nm were calculated for glass in a moist environment, with the size increasing to 25 nm in dry environments. While this model fit experimental crack growth data (Fig. 26), this fit is not enough proof that

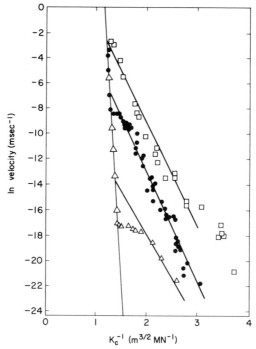

Fig. 26. Crack velocity data of soda-lime glass plotted as suggested by the plastic zone theory of Weidman and Holloway (1974).

the model is correct. In fact, plastic zone sizes calculated from the crack velocity data agreed with those expected from the flaw stress only by adjustment of a constant in one of their equations. While this model cannot be ruled out, it will take more convincing experimental evidence to show that crack growth rates in glasses are really governed by plastic flow.

A recent theory of fracture (Fuller and Thomson, 1978a) focuses on the effect of the periodic nature of the bonding in a material on crack growth behavior. As shown in Fig. 27, the model was developed based on a one-

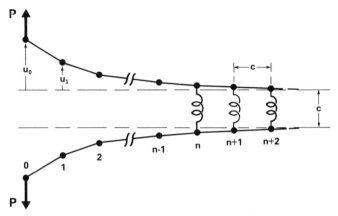

FIG. 27. Spring model of a crack in a lattice.

dimensional model of a crack, in which the lattice consisted of two semi-infinite chains of atoms. The atomic forces are modeled as flexural or tensile springs. This model proposes that in addition to the forces of the bonds at the crack tip there is a lattice-restoring force that resists crack motion. The lattice force is shown to be dependent on the elastic properties of the material.

The existence of the lattice forces leads to a potential energy function for crack motion such as illustrated in Fig. 28. The overall curve is the same as

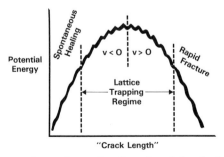

FIG. 28. Potential energy of a crack as a function of crack length showing distance over which the crack is "trapped" in the lattice.

that derived by Griffith in his original formulation for crack growth. In fact, the Griffith condition for failure is shown as the dotted line at the peak of the curve. It should be pointed out that this peak does not correspond to K_{Ic}, but is the point at which the crack energy is greater than the thermodynamic surface energy. This model explains why fracture energies are significantly higher than the surface energy. The ripples on the curve result from the lattice forces, which result in the crack being trapped over the range shown. When the crack is at the bottom of one of the small potential wells, energy is required to overcome the barrier. To the right of the curve, the smaller barriers are on the side of longer cracks, so growth occurs; to the left, crack healing is favored. One can see that at the extremes of the lattice-trapping regime the barriers disappear; a crack exceeding the final barrier incurs no further resistance to either rapid fracture or spontaneous healing.

The type of potential energy function just discussed leads to the crack velocity $-G$(or K_I) plot shown in Fig. 29. Subcritical crack growth is

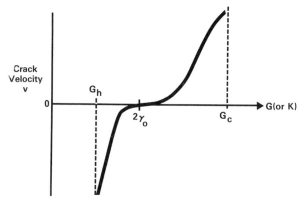

FIG. 29. Schematic of a velocity–G_I (or K_I) plot showing critical point for failure (G_c) and healing (G_h). The thermodynamic surface energy γ_0 is contained between these points.

allowed, even though no environmental effects have yet been introduced. This type of behavior may explain the crack growth data obtained in vacuum by Wiederhorn *et al.* (1974b). Also note that negative crack velocities, i.e., crack healing, are allowed by the model; crack healing has been observed in glasses. The thermodynamic surface energy γ_0 lies between the healing and growth regimes and represents a stress corrosion limit.

Recent analysis by Fuller and Thomson (1978b) has shown that chemical effects can be included in the general theory of lattice trapping. Crack growth behavior under stress corrosion conditions may then be described in terms of the effect of the chemical reaction on the atomic bonds at the crack tip.

D. STRESSING RATE TECHNIQUE

The previous sections were concerned with crack growth data obtained by fracture mechanics techniques. Crack velocities were measured directly or determined through changes in specimen compliance. It was shown that the data could be fit to an equation of the form

$$V = AK_I^n. \tag{27}$$

An alternative approach was suggested by Charles (1958), who concluded that if subcritical crack growth takes place, the strength of a material σ at a constant temperature should vary with stressing rate $\dot{\sigma}$ according to the following expression,

$$\sigma = B\dot{\sigma}^{(1/n+1)}, \tag{28}$$

where B is a constant.

A similar expression can be derived from fracture mechanics expressions (Evans, 1974; Wiederhorn, 1974):

$$d\sigma = \dot{\sigma}\, dt \tag{29}$$

and

$$d\sigma/da = \dot{\sigma}\, dt/da = \dot{\sigma}/V. \tag{30}$$

Substituting the expressions for V given in Eq. (27) and rearranging,

$$d\sigma = \dot{\sigma}\, da/AK_I^n, \tag{31}$$

since

$$K_1 = Y\sigma a^{1/2},$$

then

$$\sigma^n\, d\sigma = \dot{\sigma}\, da/AY^n a^{n/2} \tag{32}$$

and integrating over the extent of crack growth,

$$\int_0^\sigma \sigma^n d\sigma = \frac{\dot{\sigma}}{AY^n} \int_{a_i}^{a_c} a^{-n/2} da \tag{33}$$

$$\sigma^{n+1}/(n+1) = \frac{2\dot{\sigma}}{AY^2(2-n)} (a_c^{(2-n)/2} - a_i^{(2-n)/2}). \tag{34}$$

TABLE VI

Comparison of n Values Obtained from Crack Velocity and Stressing Rate Experiments

Glass	Test type[a]	Surface condition	Test environment[b]	n	Reference
Soda-lime	1	Abraded	H_2O	12–14	Mould and Southwick (1959 a,b
	1	Acid etch	50% RH	26.8	Ritter and Sherbourne (1971)
			100% RH		
	2	Acid etch and abraded	50% RH	13.0	Ritter (1969)
	2	Abraded	50% RH	16.0	Charles (1958)
	2	Abraded	6N NaOH	19.5	Ritter and LaPorte (1975)
	2	Abraded	H_2O	13.0	Ritter and LaPorte (1975)
	2	Abraded	6N HCl	25.1	Ritter and LaPorte (1975)
	2	Acid polished	6N NaOH	19.3	Ritter and LaPorte (1975)
	2	Acid polished	H_2O	16.9	Ritter and LaPorte (1975)
	2	Acid polished	6N HCl	17.8	Ritter and LaPorte (1975)
	3		H_2O	16.6	Wiederhorn and Bolz (1970)
	3		6N NaOH	19.4	Wiederhorn and Johnson (1973)
	3		6N HCl	32.0	Wiederhorn and Johnson (1973)
Borosilicate	1	Abraded	100% RH	17.0	Ritter and Manthuruthil (1973)
	1	Acid polished	100% RH	21.4	Ritter and Manthuruthil (1973)
	2	Abraded	100% RH	27.4	Ritter and Sherbourne (1971)
	2	Abraded	6N NaOH	22.6	Ritter and LaPorte (1975)
	2	Abraded	H_2O	35.1	Ritter and LaPorte (1975)
	2	Abraded	6N HCl	26.9	Ritter and LaPorte (1975)
	2	Acid polished	6N HaOH	21.1	Ritter and LaPorte (1975)
	2	Acid polished	H_2O	39.8	Ritter and LaPorte (1975)
	2	Acid polished	6N HCl	64.9	Ritter and LaPorte (1975)
	3		H_2O	34.1	Ritter and LaPorte (1975)
	3		6N NaOH	22.7	Ritter and LaPorte (1975)
	3		6N HCl	57.0	Ritter and LaPorte (1975)
Fused silica	1	As-drawn fiber	50% RH	27	Proctor et al. (1967)
	2	As-drawn fiber	50% RH	32	Proctor et al. (1967)
	2	Abraded	100% RH	37.8	Ritter and Sherbourne (1971)
	3		H_2O	36.1	Wiederhorn and Bolz (1970)
	2	Plastic-clad fiber	45% RH	21.8–29.9[c]	Tariyal and Kalish (1978)
	2	Plastic-clad fiber	97% RH	15.3–25.3[c]	Tariyal and Kalish (1978)

[a] (1) Static fatigue; (2) stress rate; (3) crack velocity.
[b] All tests performed at room temperature.
[c] n value depended on type of plastic coating.

Since $a_c^{(2-n)/2} >> a_i^{(2-n)/2}$, the former can be dropped. The initial flaw size is given by

$$a_i^{1/2} = K_{Ic}/Y\sigma_{Ic}, \tag{35}$$

where σ_{Ic} is the strength in an environment in which no subcritical crack growth will occur; then,

$$\sigma^{n+1} = \frac{2(n + 1)\,\dot\sigma\,\sigma_{Ic}^{n-2}}{A Y^2(n - 2)K_{Ic}^{n-2}}. \tag{36}$$

A plot of $\dot\sigma$ versus σ yields a straight line of slope $1/(n + 1)$, the same as derived by Charles (1958), Evans (1974), and Davidge et al. (1973).

The development of the stressing rate technique for obtaining crack growth data was particularly important, since it allows the pertinent fracture mechanics parameters to be obtained from flaws of the same size and shape as would be present in an actual structure. A question that arises is whether the crack growth parameters measured in a stressing rate test are always the same as those obtained from crack velocity experiments. As seen in Table VI, there can be a wide variance in the value of n depending on the type of test, the surface conditions of the glass, and the test environment. Work by Ritter and Sherbourne (1971) initially suggested that n values obtained by crack velocity or stressing rate techniques are equivalent within experimental error. Similarly, work by Wiederhorn et al. (1974a) indicated the equivalence of crack velocity and stressing rate as shown by Fig. 30. However, it was later shown by Ritter and LaPorte (1975) that,

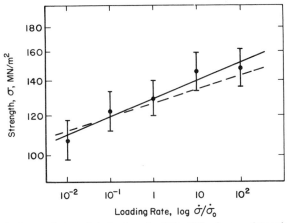

FIG. 30. Equivalence of strength behavior of a SiO_2–TiO_2 glass as determined by stressing rate experiments (solid line) or crack velocity measurements (dashed line). (After Wiederhorn et al., 1974a.)

while stressing rate and crack velocity data for soda-lime and borosilicate glasses agreed when tests were conducted in water or in NaOH, data obtained in $6N$ HC1 by the two techniques did not agree. The stressing rate tests yielded much higher n values than were obtained by a fracture mechanics technique. The authors suggested that failure under these contions involved a process that included mechanisms other than crack propagation due to stress corrosion.

A good example of the possible problems that can arise is given by the data for high-strength glass fibers that exhibit quite different stress corrosion parameters than those determined from bulk materials of the same composition. Measurements made on fused silica, the basic component of the fiber, yield an n value of ~35 (Table VI). Stressing rate experiments performed in 50% RH air on high-strength SiO_2 fibers, however, have yielded an $n \sim 22$–24 (Table VI). There are indications that if the stressing rate experiments are performed in H_2O, n is even lower, ~15. Such a change in n with variations in relative humidity is not easily explainable by current stress corrosion theories.

It should be pointed out, however, that recent work by Jakus *et al.* (1979) has indicated that large variations in n can be expected due to sampling uncertainties. They suggest that at least 200 specimens must be used in the determination of n in order to reduce scatter sufficiently.

Metcalf and Schmitz (1972) showed that the delayed failure of E glass fibers in H_2O was likely due to the exchange of hydrogen for Na^+ at the glass surface, and that strength losses were much greater in acidic solutions. In some cases, spontaneous cracking of fibers occurred, apparently due to stresses produced by the smaller hydrogen ions introduced into the structure in place of larger Na^+ ions.

E. DELAYED FAILURE

The previous section described the phenomenon of subcritical crack growth and discussed possible mechanisms that could cause it. This section will discuss the implications of subcritical crack growth for failure and show how crack growth data can be used in structural design. The phenomenon of delayed failure, or static fatigue, in which glass fails after some period of time under stress has been recognized for many years. It will be the purpose of this section to treat this phenomenon from a fracture mechanics point of view and to show how fracture mechanics concepts can be used to explain and predict the occurrence of time-dependent failure in glasses.

Although previous workers had studied the problem of delayed failure (e.g., Holland and Turner, 1940; Baker and Preston, 1946; Gurney and Pearson, 1949) and derived a number of models to explain it, the first really comprehensive study of this phenomenon was reported by Mould and

Southwick (1959a,b). Mould and Southwick showed that for a given glass composition, time-to-failure data as a function of stress could be plotted so that data points for specimens having different surface treatments all fell on the same curve. This "universal fatigue curve" as they called it was arrived at by normalizing the applied stress by the strength of the same set of specimens in liquid nitrogen and normalizing the times by the time-to-failure at 50% of the liquid nitrogen strength, leading to the plot shown in Fig. 31.

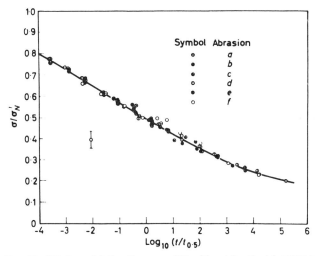

FIG. 31. "Universal fatigue" curve of Mould and Southwick (1959b).

Mould and Southwick (1959b) also showed empirically that $t_{0.5}$ should be related to the reciprocal of the square of the liquid nitrogen strength as shown in Fig. 32. The data in this figure were replotted as a logarithmic function of $t_{0.5}$ on σ_n (after Wiederhorn and Bolz, 1970), rather than the original linear plot for ease of comparison to theory. Mould and Southwick suggested that the two curves in Fig. 32 represent the data for two different types of surface flaws. The upper curve represents specimens containing what Mould and Southwick referred to as point flaws (grit blasted or ground perpendicular to the stress) while the lower curve represents specimens containing linear flaws (ground parallel to the stress). Wiederhorn and Bolz (1970) showed that the relationship between $t_{0.5}$ and $1/\sigma_N{}^2$ could be derived from fracture mechanics principles, leading to an expression of the form

$$t_{0.5} = Y(RTK_{Ic}/bV_{0.5})\,\sigma_N^{-2}, \tag{37}$$

where $V_{0.5}$ is the crack velocity at $\dfrac{K_{Ic}}{2}$ and b the stress intensity coefficient in the Hillig and Charles model. At a given temperature, the parameters

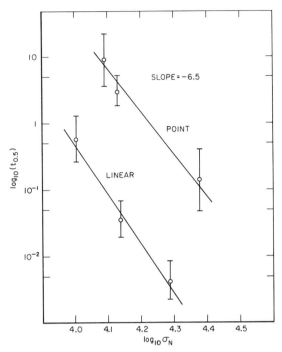

FIG. 32. Effect of flaw geometry on the time-to-failure of soda-lime glass. Plot of Wieder-horn and Bolz (1970) based on data of Mould and Southwick (1959b). Fracture mechanics theory indicates slope should be −2.

within the parentheses are constant, so that a plot of log $t_{0.5}$ versus log σ_N should yield straight lines of slope −2, separated in time by a distance Y, where Y depends on the flaw geometry. Y equals 1.25 for a halfpenny crack and 0.5 for an infinitely long-slit crack. Although Wiederhorn and Bolz found that the separation of the two curves was approximately that predicted by fracture mechanics theory, the slopes of the curves were about equal to −6.5 rather than −2. No explanation for this inconsistency was given, but the problem may lie in the way in which Mould and Southwick (1959a) obtained their data: they loaded specimens for short intervals. If the specimen did not fail within this period of time, it was reloaded at a higher stress, and so on. Mould and Southwick claimed that if one-half of the load was subtracted from each value, results would be comparable to those specimens that had experienced no prior loading. However, they did not justify this either mathematically or experimentally. Whether errors introduced by this procedure are the cause for the discrepancy between theory and experiment or there is a more fundamental cause for this difference is not known.

Other workers have attempted to compare the universal fatigue curve for various glass compositions. Ritter and Manthuruthil (1973) showed that the delayed failure curves for a number of glasses were quite sensitive to surface finish (Table VI). This result would not be predicted from the model of Mould and Southwick, since their universal fatigue curve should vary only with glass composition.

F. FAILURE PREDICTION

As just discussed, delayed failure of a ceramic component under load must always be considered as part of the total design of the structure. One of the most important aspects of the work on subcritical crack growth is the development of schemes for prediction of safe lifetimes. These lifetime prediction techniques are all derived on the basis that the subcritical crack growth can be characterized by an analytic function. The functional form most commonly used is that given in Eq. (27). However, it is strictly an empirical expression and may in fact not be the best fit to the data (Wiederhorn, 1977). However, because no theoretical model is available and because expressions for failure time are derivable from this expression (this is not necessarily the case for others), it will be used in the following derivation. Equation (27) can be rearranged such that

$$dt = \frac{da}{AK_I^n}. \tag{38}$$

Using the general expression for K_I as a function of flaw size [Eq. (16)] at a constant applied stress σ_a, one can write

$$da = \frac{2K_I \, dK_I}{Y^2 \sigma_a^2}, \tag{39}$$

$$\int_0^{t_f} dt = \frac{2}{A Y^2 \sigma_a^2} \int_{K_{Ii}}^{K_{Ic}} K_I^{1-n} \, dK, \tag{40}$$

where t_f is the time to failure and K_{Ii} is the stress intensity at the largest initial flaw. Performing the integration,

$$t_f = \frac{2}{A Y^2 \sigma_a^2 (2 - n)} [K_{Ic}^{2-n} - K_{Ii}^{2-n}]. \tag{41}$$

Since $K_{Ic}^{2-n} \ll K_{Ii}^{2-n}$ for large ns, to a good approximation,

$$t_f = 2K_{Ii}^{2-n}/A Y^2 \sigma_a^2 \, (n - 2). \tag{42}$$

Knowledge of K_{Ii}, which means knowledge of the initial flaw size distribution, is needed to allow us to predict failure times. This knowledge can be obtained in three ways: (1) nondestructive evaluation (NDE), (2) statistics;

and (3) proof testing. While great strides are being made in the development of more-sensitive NDE techniques, there is currently no method that can accurately detect surface cracks in glass in the necessary size range (10–50 μm). This leaves the latter two techniques to determine initial flaw sizes.

The statistical technique follows the approach developed by Weibull (1939). He showed that the fracture of the specimen population P that would break at a stress σ_{Ic} is given by

$$P = 1 - \exp\left[-(\sigma_{\text{Ic}} - \sigma_1)/\sigma_0\right]^m, \qquad (43)$$

where σ_1, σ_0, and m are empirical constants, and σ_{Ic} is the strength of the glass in an inert environment, i.e., where no subcritical crack growth will occur. The slope of a plot of log $[1/(1 - P)]$ versus log σ_{Ic} is given by m and the intercept by m log σ_0. σ_1, the lowest possible strength, is usually taken to be zero. One can show that

$$\sigma_{\text{Ic}} = (K_{\text{Ic}}/K_{\text{Ii}}) \, \sigma_{\text{a}}. \qquad (44)$$

Substituting Eq. (44) into Eq. (43) yields

$$K_{\text{Ii}} = K_{\text{Ic}} \left(\frac{\sigma_{\text{a}}}{\sigma_0}\right) \left[\log\left(\frac{1}{1 - P}\right)\right]^{-1/m}. \qquad (45)$$

This expression can be substituted into Eq. (42) to obtain a relation between time-to-failure and failure probability. At any failure probability P,

$$t_{\text{f}} \propto \sigma_{\text{a}}^{-n}. \qquad (46)$$

Plots of t_{f} versus log σ_{a} are a series of straight lines of slope $-n$ (Fig. 33). While such a plot is quite similar to that used for static fatigue data (Fig. 32), it is more general, since the latter represents only average failure strength. Through the use of an expression such as Eq. (46), if one knows the service stress, the probability of failure can be determined from the service stress for any desired lifetime.

Caution must always be used in projecting data obtained from small laboratory specimens to large components since the probability of failure increases with increasing size, e.g., surface area under stress. One must also consider the possibility of variations in m and σ_0. Possible errors in the determination of all of the measured parameters must be taken into account in designing the component. Various techniques in estimating these errors have been proposed (Wiederhorn et al., 1976; Jacobs and Ritter, 1976).

Proof testing can eliminate much of the uncertainty associated with the statistical approach. This involves the loading of a component to a stress higher than the known service stress. All components having flaws larger than a critical flaw size will fail during the proof test. Provided that the

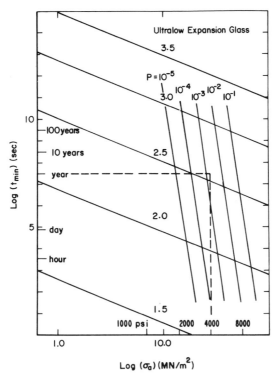

FIG. 33. Design diagram. Nearly vertical lines represent probabilities of failure based on a Weibull analysis. More-shallow lines represent different ratios of proof stress to service stress.

proof test is carried out correctly, the remaining component will have a maximum flaw size, such that

$$K_{Ii} < K_{Ic} \sigma_a / \sigma_P. \tag{47}$$

This value for K_{Ii} can be used in Eq. (42) to give a minimum time for failure at an applied stress σ_a (Fig. 33). In order to ensure reliability to longer times, the proof test ratio can be increased, but at a cost of more specimens broken during proof test. There are always uncertainties in the fracture mechanics data used to generate the design diagram, thus leading to uncertainties in the predicted lifetimes. These uncertainties in reliability can be reduced by increasing the proof stress by a specified amount.

VI. Thermal Shock

Because of the absence of any stress-relieving mechanism, high stresses can be developed in glasses due to a rapid change in temperature. These stresses act on flaws in the glass in much the same way as mechanical loading.

Although Kingery (1955) discussed the concept of strain energy release associated with crack growth during thermal shock, Hasselman (1963) was the first to show how thermal fracture could be treated from a fracture mechanics viewpoint. He showed that the degree of damage that occurred during thermal shock was a function of the amount of strain energy that could be absorbed by the material during crack propagation. He derived two parameters that could be used to predict the relative degree of damage taking place in materials, namely,

$$R''' = E/\sigma^2 \, (1 - \nu), \tag{48}$$

and

$$R'''' = E\gamma_c/\sigma^2(1 - \nu) \tag{49}$$

R''' represents the condition in which there is a minimum in the elastic energy available for crack growth; R'''' is the parameter representing a minimum in the extent of crack propagation.

One should realize that the above parameters say nothing about the severity of the thermal shock that a material can withstand without incurring some damage. The ΔT through which a material can be quenched before a decrease in strength takes place is given by other thermal shock resistance parameters. These parameters differ depending on the heat flux. At high rates of heating, the thermal stresses are generated so quickly that the effect of the thermal conductivity of the materials in dissipating the heat is minimal. The thermal stress resistance parameter that is presumed to apply under these conditions is

$$R = \sigma(1 - \nu)/\alpha E, \tag{50}$$

where α is the thermal expansion coefficient over the pertinent temperature range. For milder heat flux conditions, the thermal conductivity of the material k does play a part in determining thermal shock resistance, leading to

$$R' = \sigma(1 - \nu)k/\alpha E. \tag{51}$$

Following Hasselman (1970), the use of these parameters in predicting thermal shock behavior is shown schematically in Fig. 34.

It is apparent from Eqs. (50) and (51) that it is the strength of a material rather than simply its fracture toughness that will determine its thermal shock resistance. That is, $YK_{Ic}/\sqrt{a}$ must be considered. Thermal shock resistance parameters based on strength, such as R and R', are even too simple for adequately ranking materials as to their thermal shock resistance. One of the major reasons these thermal stress resistance parameters

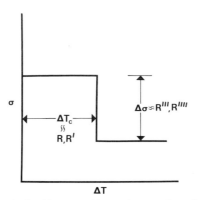

FIG. 34. Typical curve obtained by quenching a glass specimen in water from an elevated temperature showing associated thermal stress ranking parameters.

are not accurate is that the strength value used in the expression is usually the flexural strength of the material. The biaxial stress field, which occurs during thermal shock, leads to much lower values in strength and can interact with an entirely different set of flaws. Experimentally, it has been observed that differences in flexural strength of ~50% lead to almost no change in ΔT_c for a water quench. This problem was recognized by Hasselman et al. (1976), who found that the flaw size used to predict thermal stress failures was approximately a factor of 3 greater than that calculated from a flexural strength data. In a later study, Hasselman et al. (1978a) observed that if large flaws were introduced into glass rods using a hardness indentor, thermal stress failures could be quite accurately predicted. The reason for the good agreement was the fact that the use of large artificial flaws eliminated the statistical effects usually encountered with brittle materials.

Badaliance et al. (1974) showed that another important factor in predicting the thermal shock behavior of glass is subcritical crack growth. These authors used an expression for the transient thermal stresses at the surface of a cylinder and combined this expression with the formulation for crack velocity as a function of K_I given in Eq. (27). They showed that one could derive an expression for crack length as a function of time, i.e.,

$$a(t) = a_0 + \int_0^t V \, dt. \tag{52}$$

The above equations were used to predict the ΔT_c of soda-lime glass rods quenched into a water bath. The crack growth data fed into these expressions were obtained on a similar glass by Wiederhorn and Bolz (1970). The essential results of this study are shown in Fig. 35 as a plot of ΔT_c versus the Biot modulus for different flaw sizes. Biot's modulus m is a measure of

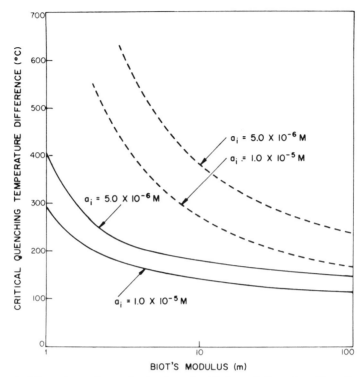

FIG. 35. Effect of quench severity ($\approx m$) on the ΔT_c for soda-lime glass. The effect of subcritical crack growth on ΔT_c is shown for two different flaw sizes. ----, No crack growth,; ——, with crack growth. (After Badaliance *et al.*, 1974.)

the severity of the thermal shock where

$$m = \frac{\text{rod radius} \times \text{heat transfer coefficient}}{\text{thermal conductivity of the glass}}.$$

It can be seen that if subcritical crack growth is not taken into account, the ΔT through which the glass can be quenched without degrading the strength is significantly overestimated. The fact that the use of the subcritical crack growth data allows one to accurately predict ΔT_c is shown in Fig. 36, which compares the predicted and measured values of strength as a function of ΔT.

VII. Impact and Erosion

Of significant recent interest is the damage caused to brittle materials by impact phenomena. Impact damage is important not only from the point of

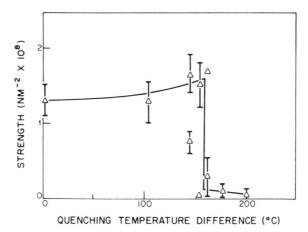

FIG. 36. Predicted (curve) versus actual (data points) thermal stress resistance of soda-lime glass. Agreement between data and prediction is indicative of the necessity of taking into account subcritical crack growth. (After Badaliance *et al.*, 1974.)

view of strength degradation and erosion, but also because the ceramic machining process can be described in terms of impact.

A. STATIC INDENTATIONS

While there are clearly dynamic effects involved in impact, it is first beneficial to examine the effect of static contact loads on surface damage. Indentations can be separated into essentially two classes, those caused by "blunt" indentors and those caused by "sharp" indentors, depending on whether contact is governed by elastic or plastic conditions. In the case of the blunt indentor, typically a sphere, crack initiation follows the sequence shown in Fig. 37 (Lawn and Marshall, 1978). In this case, existing surface flaws are the source of the eventual cone crack that forms during indentation. Frank and Lawn (1967) showed that the load P_c required to initiate the cone crack is given by

$$P_c = \alpha r K_c^2 / E, \tag{53}$$

where r is the radius of the indenting sphere and α a dimensionless constant that depends on the materials' properties of the specimen and the indentor. The applicability of Eq. (53) and the fact that the critical load is independent of the initial flaw size in the specimen surface has been confirmed experimentally in a study performed on soda-lime glass (Langitan and Lawn, 1969). Equation (53) tends to break down for small flaws and large spheres and for conditions where there are few surface flaws present (Lawn and Marshall, 1978).

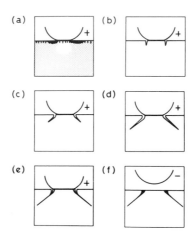

FIG. 37. Evaluation of cone crack pattern; (a) blunt indentor subjects surface flaws to tensile stress, (b) ring crack forms and (c), (d) grows until (e) cone is formed; (f) cone crack tries to close during unloading, but cannot because of debris in the crack. (After Lawn and Marshall, 1978.)

Once the typical cone crack develops, further crack growth no longer depends on the initiation site. Roesler (1956) and later Lawn and Wilshaw (1975) showed that under these conditions the size of the cone crack was related to the indentation load by

$$P/c^{3/2} = \beta K_c,\qquad(54)$$

where β is a dimensionless constant dependent on the specimen material. This relationship was also shown to be valid for the formation of radial cracks (Lawn and Fuller, 1975).

In the case of sharp indentations produced by Knoop or Vickers indentors, nonelastic processes (e.g., plastic flow, densification) at the indentor site become important. The sequence of events leading to cracking under these conditions is shown schematically in Fig. 38 (Lawn and Swain, 1975). Two sets of cracks are formed, "median" or "radial" cracks, which form as semicircles perpendicular to the surface and lateral cracks that initiate below the deformation zone, then grow outward and upward until they reach the surface. The first set of cracks causes strength degradation, while the second set is responsible for loss of material during erosion.

Lawn and Evans (1978) derived an expression for the critical load needed to form the median cracks:

$$P_c = \alpha_p K_c^4/H^3,\qquad(55)$$

where H is the hardness of the material and α_p a dimensionless constant that is a function of the elastic–plastic zone parameters in the material. As

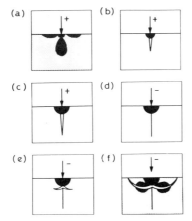

FIG. 38. Evolution of median and lateral cracks during loading with a sharp indentor: (a) stress (dark area) occurs under indentor, (b), (c) subsurface flaws grow to form median cracks, (d) during unloading lateral cracks form and (e), (f) expand and grow to the surface. (After Lawn and Marshall, 1978.)

in the case of the blunt indentor, this expression does not involve the initial flaw size in the specimen. It is assumed that median cracks form from flaws in the elastic–plastic zone, but the region of these flaws is not clear. Although Eq. (55) has not been completely verified, it does fit the general trend of experimental results (Evans and Wilshaw, 1976). Veldkamp *et al.* (1978) derived the same expression for sliding indentation.

Lateral cracking is a more complicated process. While some analysis has been performed relating lateral crack sizes to the median crack size, the use of fracture mechanics to understand this phenomenon has not been well developed.

B. DYNAMIC LOADING

Conditions for crack formation during an actual impact can be derived from our knowledge of indentation fracture. Impacting particles can be characterized as being blunt or sharp. By assuming that all of the kinetic energy of a blunt particle is converted into elastic energy during impact, Evans (1973) and Wiederhorn and Lawn (1979) derived an expression for the impulse load F_m imported to the target

$$F_m = 2.4(E/k)^{2/5}\rho^{3/5}r^2V_0^{6/5}, \qquad (56)$$

where ρ is the particle density, r its radius, V_0 its velocity, and k a constant that depends on the elastic properties of the particle and the target. A critical velocity for crack formation can be obtained by combining Eq. (56)

with an expression for Auerbach's law, yielding Eq. (53), then (Wiederhorn and Lawn, 1977),

$$V_c = 0.48(k/E)^{7/6} K_c^{5/3}/\rho^{1/2} r^{5/6} \phi^{*5/6} \tag{57}$$

Brittle materials impacted by particles having velocities greater than V_c will exhibit decreases in strength and loss of material by erosion. The strength degradation that results from an impact was shown by Wiederhorn and Lawn (1979) to be given by

$$\sigma = [0.42(k/E)^{2/15}\beta_R^{1/3} K_c^{4/3}/\Omega^{1/2}\rho^{1/5} r^{2/3}] V_0^{-2/5}, \tag{58}$$

where β_R and Ω are constants and the other terms have been previously defined. Wiederhorn and Lawn showed that the above expression predicted the strength behavior of glass impacted with steel and tungsten carbide spheres as shown in Fig. 39.

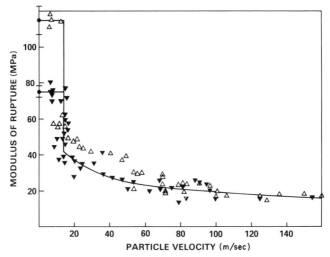

FIG. 39. Strength degradation due to particle impact on soda-lime glass. Solid curve is a representation of Eq. (58) and is seen to fit the experimental data quite well. Steel spheres: 0.8 mm; ▼, 100-grit abraded glass; Δ, 320-grit abraded glass. (After Wiederhorn and Lawn, 1979.)

Because plastic deformation occurs, impact by sharp indentors is a more complicated process. Lawn and Evans (1977) assumed that cracking occurs from preexisting flaws due to the stresses set up by the plastic zone. They derived expressions from fracture mechanics theory, which relate the critical flaw size C^* and critical load F^* for fracture as a function of the hardness and K_{Ic} of the target:

$$C^* = 1.767 K_c^2/\theta^2 H,$$
$$F^* = (54.57\alpha K_c^3/E^2\theta^4 H)K_c, \tag{59}$$

where α, θ, and ξ are constants. Lawn and Evans (1977) calculated that the critical load for fracture of glass from a sharp indentor was $0.02N$, much less than that required for impact of a spherical particle (e.g., 0.4-mm-radius particle requires $98.5N$).

One of the major questions that must be answered is how the kinetic energy of the particle is transferred to the target. Two models have been suggested for determining the contact force. Wiederhorn and Lawn (1979) calculated the contact force from the hardness of the target and the maximum depth of penetration of the particle by assuming that all of the particle's kinetic energy is dissipated in plastic flow in the target:

$$F \propto H^{1/3}m^{2/3}V_0^{4/3}, \qquad (60)$$

where m is the mass of the particle. The crack size formed during the impact is given by

$$C_r \propto m^{4/9}H^{2/9}V_0^{8/9}K_c^{-2/3}. \qquad (61)$$

Evans *et al.* (1978) included dynamic effects in their impact model. The contact force was calculated from the dynamic pressure arising when the particle first hit the surface. The depth of penetration was determined from the time of contact and the mean interface velocity. Their expressions for the contact force F_m and crack size formed C_r are given by

$$F_m \propto V^2r^2\rho, \qquad (62)$$
$$C_r \propto [(Vr^2)/K_c]^{2/3}\rho^{2/3}, \qquad (63)$$

where r is the particle radius and ρ its density.

The two theories suggest different dependencies of crack formation on particle velocity and the hardness of the target. In general, reasonable agreement with experiment has been claimed for both theories. Further work is required to establish the superiority of one model over the other.

C. EROSION

Actual erosion damage to ceramics results from a large number of particle impacts and can be a complicated process. Two models of erosion of brittle ceramics have been developed. Sheldon and Finnie (1966a,b) assumed that erosion is due to Hertzian contact stresses during impact. These stresses cause preexisting cracks to grow, so that the load to cause crack propagation can be related to the size distribution of surface flaws. Their expression for erosion rate W is

$$W = k_1r^aV^b, \qquad (64)$$

where r is the size of the eroding particle and V its velocity; a, b, and k, are functions of the Weibull modulus m and the elastic modulus and density of the particle. While reasonable fit of experimental data to this model was obtained, the theory itself has been questioned because it assumes that Hertzian cracks cause erosion. Experimental evidence indicates that lateral cracks are the major cause of material removal.

Evans *et al.* (1978) derived an erosion model by assuming that erosion rate is proportional to the amount of material removed by each impact event. This loss per impact is dependent on the lateral crack formation. Evans *et al.* derived the following relationship between erosion rate and the various properties of the particle and the target:

$$V \propto V_0^{19/6} \, r^{11/3} \rho^{19/12} K_c^{-4/3} H^{-1/4}. \tag{65}$$

More work will be required for the actual erosion mechanism to be accurately determined.

D. MACHINING

The machining of glass involves abrading the surface with particles of a harder material such as SiC or diamond. This abrasion produces scratches on the glass surface, the severity of which depends on the size of the grit particle, the forces used, and the grinding speed. Observations by Swain (1968) indicate that the cracking that occurs about the scratches is quite similar to that occurring beneath pointed indentors. Swain showed that fracture mechanics could be applied to understanding the size of the cracks formed. As noted earlier, Mecholsky *et al.* (1977) showed that these cracks are the sources of failure in the glass. Swain also showed that the residual stresses associated with the scratches had a significant effect on the strength of the glass, and that these stresses apparently decrease with time. Veldkamp *et al.* (1978) performed detailed experiments on the scratching of glass and other ceramics. They showed that Eq. (55) developed originally for static indentations could be used to predict the onset of cracking due to scratching.

VIII. Summary

The objective of this chapter was to show the valuable contribution that fracture mechanics has made to our understanding of the mechanical properties of glass. Much still remains to be learned, however, especially with regard to the effects of complex stress states, impact processes, and thermal stresses. Our knowledge of crack growth mechanisms is quite meager and must be expanded in order to provide a basis for material selection as well as for lifetime predictions. If the next ten years bring as much gain in knowledge as the last ten, we shall be well on our way to attaining many of these goals.

ACKNOWLEDGMENT

Helpful discussions with Sheldon Wiederhorn are gratefully acknowledged.

References

Abdel-Latif, A.I.A., Bradt, R. C., and Rindone, G. E. (1976). *J. Am. Ceram. Soc.* **59**, 174–175.
Abdel-Latif, A.I.A., Tressler, R. E., and Bradt, R. C. (1977). *Int. J. Fract. Mech.* **13**, 349–359.
Badaliance, R., Krohn, D. A., and Hasselman, D. P. H. (1974). *J. Am. Ceram. Soc.* **57**, 432–436.
Baker, T. C., and Preston, F. W. (1946). *J. Appl. Phys.* **17**, 170–178.
Bansal, G. K. (1978). Private communication.
Bansal, G. K., and Duckworth, W. H. (1977). *J. Am. Ceram. Soc.* **60**, 304–310.
Brown, S. D. (1978). "Fracture Mechanics of Ceramics" (R. C. Bradt, D. P. H. Hasselman, and F. F. Lange, eds.), Vol. 4, pp. 597–621. Plenum Press, New York.
Chandon, H. C., Bradt, R. C., and Rindone, G. E. (1978). *J. Am. Ceram. Soc.* **61**, 207–210.
Charles, R. J. (1958). *J. Appl. Phys.* **29**, 1657–1662.
Clarke, A. B. J., and Irwin, G. R. (1966). *Exp. Mech.* **23**, 321–330.
Cox, S. M. (1969). *Phys. Chem. Glasses* **10**, 226–239.
Congelton, J., and Petch, N. J. (1967). *Philos. Mag.* **16**, 749.
Davidge, R. W., McLaren, J. R., and Tappin, G. (1973). *J. Mater. Sci.* **8**, 1699–1705.
Ernsberger, F. M. (1969). *Phys. Chem. Glasses* **10**, 240–245.
Evans, A. G. (1973). *J. Am. Ceram. Soc.* **56**, 405–409.
Evans, A. G. (1974). *Int. J. Fract.* **10**, 251–259.
Evans, A. G. (1979). *Proc. Nat. Fracture Mech. Conf., 11th* VPI (to be published).
Evans, A. G., and Wilshaw, T. R. (1976). *Acta Metall.* **24**, 939–956.
Evans, A. G., Gulden, M. E., and Rosenblatt, M. E. (1978). *Proc. R. Soc. London Ser. A* **361**, 343–365.
Frank, F. C., and Lawn, B. R. (1967). *Proc. R. Soc. London Ser. A* **299**, 291–306.
Freiman, S. W. (1974). *J. Am. Ceram. Soc.* **57**, 350–353.
Freiman, S. W. (1975). *J. Am. Ceram. Soc.* **58**, 340–341.
Freiman, S. W., Mulville, D. R., and Mast, P. W. (1973). *J. Mater. Sci.* **8**, 1527–1534.
Freiman, S. W., Gonzalez, A. C., and Mecholsky, J. J. (1979). *J. Am. Ceram. Soc.* **62**, 206–208.
Fuller, E. R. (1979). *Proc. Nat. Fracture Mech. Conf., 11th* VPI (to be published).
Fuller, E. R., and Thomson, R. M. (1978a). "Fracture Mechanics of Ceramics" (R. C. Bradt, D. P. H. Hasselman, and F. F. Lange, eds.), Vol. 4, pp. 507–548. Plenum Press, New York.
Fuller, E. R., and Thomson, R. M. (1978b). Private communication.
Gerberich, W. W., and Stout, M. (1976). *J. Am. Ceram. Soc.* **58**, 222–225.
Griffith, A. A. (1921). *Trans. R. Soc. London* **221**, 163–198.
Gunsakera, S. P., and Holloway, D. G. (1973). *Phys. Chem. Glasses* **14**, 45–52.
Gurney, C., and Pearson, S. (1949). *Proc. Phys. Soc. London* **62**, 469–476.
Hasselman, D. P. H. (1963). *J. Am. Ceram. Soc.* **46**, 535–540.
Hasselman, D. P. H. (1970). "Ultrafine-Grain Ceramics" (J. J. Burke, N. L. Reed, and V. Weiss, eds.), pp. 297–315. Syracuse Univ. Press, Syracuse, New York.
Hasselman, D. P. H., Badaliance, R., McKinney, K. R., and Kim, C. H. (1976). *J. Mater. Sci.* **11**, 458–464.

Hasselman, D. P. H., Chen, E. P., and Urick, P. A. (1978a). *Bull. Am. Ceram. Soc.* **57**, 432–436.

Hasselman, D. P. H., Chen, E. P., and Urick, P. A. (1978b). *Bull. Am. Ceram. Soc.* **57**, 189–192.

Hillig, W. B., and Charles, R. J. (1965). "High Strength Materials" (V. F. Zackay, ed.), pp. 682–705. Wiley, New York.

Holland, A. J., and Turner, W. E. S. (1940). *J. Soc. Glass Technol.* **24**, 46–57T.

Hussian, M. A., Pu, S. L., and Underwood, J. (1974). ASTM-STP No. 381, pp. 2–28. American Society of Testing and Materials, Philadelphia, Pennsylvania.

Inglis, C. E. (1913). *Trans. Inst. Nav. Arch.* **55**, 219–229.

Irwin, G. R. (1958). "Encyclopedia of Physics," Vol. VI, pp. 551–590, Springer, Berlin.

Jacobs, D. F., and Ritter, J. E., Jr. (1976). *J. Am. Ceram. Soc.* **59**, 481–487.

Jakus, K., Ritter, J. E., Jr., and Bandyopadayay, N. (1979). *Bull. Am. Ceram. Soc.* **58**, 312.

Johnson, J. W., and Holloway, D. G. (1966). *Philos. Mag.* **14**, 731.-743.

Kerper, M. J., and Scuderi, T. G. (1965). *Bull. Am. Ceram. Soc.* **44**, 953–955.

Kingery, W. D. (1955). *J. Am. Ceram. Soc.* **38**, 3–15.

Kirchner, H. P. (1976). *Proc. Int. Conf. Mech. Behavior Mater., 2nd* pp. 1317–1321. American Society of Metals.

Kirchner, H. P., and Gruver, R. M. (1974). "Fracture Mechanics of Ceramics" (R. C. Bradt, D. P. H. Hasselman, and F. F. Lange, eds.), Vol. 1, pp. 309–321. Plenum Press, New York.

Kobayashi, A. S., Staley, L. I., Emory, A. F., and Love, W. J. (1978). "Fracture Mechanics of Ceramics" (R. C. Bradt, D. P. H. Hasselman, and F. F. Lange, eds.), Vol. 3, pp. 451–461. Plenum Press, New York.

Langitan, F. B., and Lawn, B. R. (1969). *J. Appl. Phys.* **39**, 4828.

Langitan, F. B., and Lawn, B. R. (1970). *J. Appl. Phys.* **41**, 3357–3365.

Lawn, B. R., and Evans, A. G. (1977). *J. Mater. Sci.* **12**, 2195–2199.

Lawn, B. R., and Evans, A. G. (1978). *J. Mater. Sci.* (to be published).

Lawn, B. R., and Fuller, E. R. (1975). *J. Mater. Sci.* **10**, 2016–2024.

Lawn, B. R., and Marshall, D. B. (1978). "Fracture Mechanics of Ceramics" (R. C. Bradt, D. P. H. Hasselman, and F. F. Lange, eds.), Vol. 3, pp. 205–229. Plenum Press, New York.

Lawn, B. R., and Swain, M. V. (1975). *J. Mater. Sci.* **10**, 113–122.

Lawn, B. R., and Wilshaw, T. R. (1975). *J. Mater. Sci.* **10**, 1049–1081.

Linger, K. R., and Holloway, D. G. (1968). *Philos. Mag.* **16**, 1269–1280.

Mecholsky, J. J., and Freiman, S. W. (1979). *ASTM-STP 678.* American Society of Testing and Materials, Philadelphia, Pennsylvania (to be published).

Mecholsky, J. J., Rice, R. W., and Freiman, S. W. (1974). *J. Am. Ceram. Soc.* **57**, 440–443.

Mecholsky, J. J., Freiman, S. W., and Rice, R. W. (1976). *J. Mater. Sci.* **11**, 1310–1319.

Mecholsky, J. J., Freiman, S. W., and Rice, R. W. (1977). *J. Am. Ceram. Soc.* **60**, 114–117.

Mecholsky, J. J., Freiman, S. W., and Rice, R. W. (1978). ASTM-STP 645, pp. 363–379. American Society of Testing and Materials, Philadelphia, Pennsylvania.

Mecholsky, J. J., Gonzalez, A. C., and Freiman, S. W., (1979). *J. Am. Ceram. Soc.* (to be published).

Metcalf, A. G., and Schmitz, G. K. (1972). *Glass Technol.* **13**, 5–16.

Mott, N. F. (1948). *Engineering* **165**, 16–18.

Mould, R. E. (1961). *J. Am. Ceram. Soc.* **44**, 48–491.

Mould, R. E., and Southwick, R. D. (1959a). *J. Am. Ceram. Soc.* **42**, 542–547.

Mould, R. E., and Southwick, R. D. (1959b). *J. Am. Ceram. Soc.* **42**, 582–592.

Outwater, J. O., and Gerry, D. J. (1966). Interim Report, Contract No. Nonr-3291(01) (x). Univ. Vermont, Burlington, Vermont.

Petrovic, J. J., and Mendiratta, M. G. (1976). *J. Am. Ceram. Soc.* **59**, 163–167.
Petrovic, J. J., and Mendiratta, M. G. (1977). *J. Am. Ceram. Soc.* **60**, 463.
Petrovic, J. J., Jacobson, L. A., Talty, P. K., and Vasuclevan, A. K. (1975). *J. Am. Ceram. Soc.* **58**, 113–116.
Proctor, B. A., Whitney, I., and Johnson, J. W. (1967). *Proc. R. Soc. London Ser. A* **297**, 534–557.
Quackenbush, C. L. W., and Frechette, V. D. (1978). *J. Am. Ceram. Soc.* **61**, 207–210.
Ritter, J. E., Jr. (1969). *J. Appl. Phys.* **40**, 340–344.
Ritter, J. E., Jr., and LaPorte, R. P. (1975). *J. Am. Ceram. Soc.* **58**, 265–267.
Ritter, J. E., Jr., and Manthuruthil, J. (1973). *Glass Technol.* **14**, 60–64.
Ritter, J. E., Jr., and Sherbourne, C. L. (1971). *J. Am. Ceram. Soc.* **54**, 601–605.
Roberts, D. K., and Wells, A. A. (1954). *Engineering* **178**, 820–821.
Roesler, F. C. (1956). *Proc. Phys. Soc. London Sect. B* **69**, 981–992.
Shand, E. B. (1954). *J. Am. Ceram. Soc.* **37**, 559–572.
Shand, E. B. (1959). *J. Am. Ceram. Soc.* **42**, 474–477.
Sheldon, G. L., and Finnie, I. (1966a). *Trans. ASME J. Eng. Ind.* **88**, 387–392.
Sheldon, G. L., and Finnie, I. (1966b). *Trans. ASME J. Eng. Ind.* **88**, 393–400.
Sneddon, I. N. (1946). *Proc. R. Soc. London Ser. A* **187**, 229–260.
Stevens, R. N., and Dutton, R. (1971). *Mater. Sci. Eng.* **8**, 220–234.
Swain, M. V. (1978). "Fracture Mechanics of Ceramics" (R. C. Bradt, D. P. H. Hasselman, and F. F. Lange, eds.), Vol. 3, pp. 257–272. Plenum Press, New York.
Tariyal, B. K., and Kalish, D. (1978). "Fracture Mechanics of Ceramics" (R. C. Bradt, D. P. H. Hasselman, and F. F. Lange, eds.), Vol. 3, pp. 161–175. Plenum Press, New York.
Thomson, R., Hsieh, C., and Rana, R. (1971). *J. Appl. Phys.* **42**, 3154–3160.
Tyson, W. R., Cekirge, H. M., and Krause, A. S. (1976). *J. Mater. Sci.* **11**, 780–784.
Veldkamp, J. D. B. Hattu, N., and Snijders, V. A. C. (1978). "Fracture Mechanics of Ceramics" (R. C. Bradt, D. P. H. Hasselman, and F. F. Lange, eds.), Vol. 3, pp. 273–301. Plenum Press, New York.
Weibull, W. (1939). *Ingenioersvetenskapsakad. Handl.* No. 151.
Weidman, G. W., and Holloway, D. G. (1974). *Phys. Chem. Glasses* **15**, 68–75.
Wiederhorn, S. M. (1967). *J. Am. Ceram. Soc.* **50**, 407–414.
Wiederhorn, S. M. (1969). *J. Am. Ceram. Soc.* **52**, 99–105.
Wiederhorn, S. M. (1972). *J. Am. Ceram. Soc.* **55**, 81–85.
Wiederhorn, S. M. (1974). "Fracture Mechanics of Ceramics" (R. C. Bradt, D. P. H. Hasselman, and F. F. Lange, eds.), Vol. 2, pp. 613–646. Plenum Press, New York.
Wiederhorn, S. M. (1977). "Fracture 1977" (D. M. R. Taplin, ed.), Vol. 3, pp. 893–901. Waterloo Univ. Press, Waterloo, Canada.
Wiederhorn, S. M. (1978). "Fracture Mechanics of Ceramics" (R. C. Bradt, D. P. H. Hasselman, and F. F. Lange, eds.), Vol. 4, pp. 549–580. Plenum Press, New York.
Wiederhorn, S. M., and Bolz, C. H. (1970). *J. Am. Ceram. Soc.* **53**, 543–548.
Wiederhorn, S. M., and Johnson, H. (1971). *J. Appl. Phys.* **42**, 681–684.
Wiederhorn, S. M., and Johnson, H. (1973). *J. Am. Ceram. Soc.* **56**, 192–197.
Wiederhorn, S. M., and Lawn, B. R. (1979). *J. Am. Ceram. Soc.* (in press).
Wiederhorn, S. M., Evans, A. G., Fuller, E. R., and Johnson, H. (1974a). *J. Am. Ceram. Soc.* **57**, 319–323.
Wiederhorn, S. M., Johnson, H., Diness, A. M., and Heuer, A. H. (1974b). *J. Am. Ceram. Soc.* **57**, 336–341.
Wiederhorn, S. M., Fuller, E. R., Mandel, J., and Evans, A. G. (1976). *J. Am. Ceram. Soc.* **59**, 403–411.
Wiederhorn, S. M. (1977). "Fracture 1977" (D.M.R. Taplin, ed.), Vol. 3, pp. 893–901. Waterloo Univ. Press, Waterloo, Canada.

Wiederhorn, S. M. (1978). "Fracture Mechanics of Ceramics" (R. C. Bradt, D. P. H. Hasselman, and F. F. Lange, eds.), Vol. 4, pp. 549–580. Plenum Press, New York.
Wilkins, B. J. S., and Dutton, R. (1976). *J. Am. Ceram. Soc.* **59**, 108–112.
Williams, D. P., and Evans, A. G. (1973). *J. Test. Eval.* **1**, 264–270.
Yoffee, E. H. (1951). *Philos. Mag.* **42**, 739–750.

CHAPTER 3

Inelastic Deformation and Fracture in Oxide, Metallic, and Polymeric Glasses

A. S. Argon

DEPARTMENT OF MECHANICAL ENGINEERING
MASSACHUSETTS INSTITUTE OF TECHNOLOGY
CAMBRIDGE, MASSACHUSETTS

I. Introduction

The description of plastic deformation and its anelastic transients in crystalline matter has occupied the attention of numerous workers over the past half century. The problem proved to have great complexities resulting from the almost limitless forms that dislocations can take in a crystal; due to the constraints imposed on their shape and path by crystal summetry and bonding; and finally due to the multiplicity of elastic and chemical interactions that they may have with each other and with other slip plane obstacles. In this arduous quest the inelastic behavior of noncrystalline materials has been given comparatively little attention. On the isolated occasions when such attention was bestowed, it took simplistic routes by classifying all

79

glassy materials at low temperatures either as brittle and therefore (by faulty logic) nondeformable, or, alternatively, when the evidence of their plastic flow was overwhelming, this was often described by extensions of viscous behavior or by means of generalized dislocations. The accumulating evidence now suggests that inelastic deformation in glassy matter is subject to far fewer complexities, and may be quite generally and compactly explained. On the other hand, premature fracture in glassy materials often imposes a very severe restriction on potential plastic behavior and has been largely responsible for their widespread association with brittleness. As in crystalline matter, glassy materials also exhibit interesting aging effects that influence their plastic resistance and affect their fracture threshold. In this chapter we shall discuss on a general basis the mechanisms of inelastic deformation and fracture of glassy materials under monotonically rising stress. Our discussion will include inorganic, metallic, and long-chain polymeric glasses and will elucidate the interplay between plastic flow and fracture. By necessity, we shall not discuss the behavior of partially glassy (therefore partially oriented, crystalline, or partially rubbery) materials, as their behavior will be that of a complex mixture of the behavior of the glassy phase with the other heterogeneities. We shall also have to leave out the cyclic deformation behavior of glasses and interactions with special atmospheres and environments but shall provide some key references to guide the interested reader in these directions.

II. Inelastic Deformation in Glasses

A. NONLOCAL NATURE OF PLASTIC FLOW

In the elastic response of solids that arises from changes in internal energy between atoms due to imposed stresses, the strain in the body is locally homogeneous over regions where the stress is homogeneous, and the relative displacements between neighboring atoms everywhere are small on the scale of the interatomic distances. In this case, continuous field notions apply to the deformations in the body. The material can be considered as a continuum on a scale somewhat larger than atomic dimensions and the deformation is termed *local*. Stresses and strains are definable at a point as long as the "point" is no smaller than several atom diameters. In crystalline matter, where plastic distortions occur by lattice translations across slip planes initiated by moving dislocations, strains can only be defined meaningfully over regions large in comparison with ranges of such lattice translations, i.e., on the scale of locality of elastic deformation, crystal plasticity is a *nonlocal* form of deformation. The deformation can be approximated as a continuously varying field only on a very coarse scale where such variations are smoothed out.

The nonlocal nature of plastic flow in crystalline materials is a direct consequence of the regular structure of the crystal that produces strong mechanical coupling between neighboring volume elements. Production of a large local shear strain between two layers of atoms by translating them over each other by an identity translation cannot be accomplished without affecting the surroundings. A dislocation loop surrounds the area and can readily propagate out if it is in excess of a required critical size in relation to the applied stress. The strong mechanical coupling usually manifests itself by a low resistance against such a propagation in comparison to the shear strength of the perfect crystal.

In glassy materials, the production of large local shear strains also requires relative displacements between neighboring atoms that are of the order of atomic dimensions. Since glassy materials are disordered on the atomic scale and contain a distribution of free volume, such local shears do not affect the surroundings as strongly as in a perfect crystal and can in many instances be locally accommodated without propagating out. That is, a disordered glassy structure is a relatively weakly coupled system of local volume elements. This permits mechanically isolated, large local shears in small individual volume elements of "loose" atomic packing. Thus, plastic flow in glassy materials is still a nonlocal form of deformation in comparison with elastic deformation but can be far more local than the corresponding form in crystalline matter involving propagating dislocations. We shall further develop these ideas in later sections.

B. COMPETITION BETWEEN PLASTIC FLOW AND BRITTLE CRACK
   PROPAGATION

Kelly et al. (1967) have discussed the cause of inherent ductility versus inherent brittleness in crystalline matter and have proposed that this differentiation is a result of a competition between cleavage and plastic blunting at the tip of an atomically sharp crack. Thus, if the ratio of the level of maximum tensile stress to the level of maximum shear stress at the tip of an atomically sharp crack under stress exceeds the ratio of the ideal tensile cohesive strength to the ideal shear strength of a crystal, the crack can advance by cleavage without initiating any plastic blunting and results in completely brittle behavior. In their evaluation of materials according to this criterion, Kelly et al. have found that while covalently bonded materials tend to be inherently brittle at low temperature without exception, the close-packed face-centered cubic metals are inherently ductile also without exception. In addition, in their theoretical modeling they observed that close-packed solids bound by a Lennard-Jones potential have the property of always being ductile as such solids have a ratio of ideal cohesive strength to ideal shear strength of 5.64—averaged over all orientations, while the ratio of the maximum tensile stress to the maximum shear stress at the tip of a crack is usually only around 2.60.

These results can serve as a very useful guide in grading the response of glassy materials. In inorganic glasses, where the known short-range order is almost the same as that in the corresponding crystalline phase, we expect that at low temperature atomically sharp cracks will continue to propagate by cleavage in tension without any plastic deformation, resulting in brittle behavior. On the other hand, in metallic glasses where the short-range arrangement between atoms is nearly the same as in a liquid of the same composition, we expect that the solid can be approximated well by a Lennard-Jones potential and that the alloy be potentially ductile in tension. These expectations are generally confirmed. Inorganic oxide glasses undergo brittle fracture in tension before any detectable plastic flow either on the macroscopic scale or on the microscopic scale as might be revealed from inspection of the fracture surfaces. In metallic glasses, on the other hand, a macroscopically brittle behavior is usually found at low temperatures, but inspection of the fracture surfaces reveals features that could have resulted only from intensive local plastic flow. Polymeric glasses represent an interesting intermediate behavior pattern. When completely disordered, we expect these materials to respond initially according to their intermolecular van der Waals interaction, which is also well modeled by a Lennard-Jones potential. Thus we expect polymeric glasses also to be potentially ductile in tension to the extent that crack propagation be accompanied by local plastic flow. Indeed, inspection of fracture surfaces in glassy polymers reveals intensively deformed layers that in many instances are known to be partly craze matter. In each of these cases of inorganic, metallic, or polymeric glasses, plastic deformation can be initiated at all temperatures either in compression or under suitable superimposed hydrostatic pressure. In the case of oxide glasses, however, where significant tensile stresses can still be generated at tips of cracks even in compression, the requirements for plastic deformation at low temperatures are most stringent and involve either very large superimposed hydrostatic pressures or require indenting very small volume elements by microhardness indenters. These conditions are achievable in the laboratory but are rarely encountered in practice. Hence, it must be concluded that even though conditions for fracture-free plastic flow are achievable in oxide glasses, these materials are in practice brittle at low temperatures. In metallic and polymeric glasses, where potential ductility is always present, plastic flow—albeit at times very heterogeneous—is readily achievable in compression and often in shear.

At temperatures becoming a significant fraction of that of glass transition, inelastic deformation becomes much more strain-rate dependent and homogeneous, resulting in elimination of brittleness. We shall demonstrate that such behavior is not the result of a new mechanism but a natural consequence of reduced deformation resistance with increasing temperature in the basic mechanism of inelastic deformation in glasses.

C. STRUCTURAL IDEALIZATIONS OF GLASSES

As with most materials, the mechanism of inelastic deformation of glasses is closely linked to their structure. The most widely used quantitative means of describing this structure in nonpolymeric glasses is by radial distribution functions (RDF) of atom positions, which was pioneered by Warren (1937; Warren *et al.*, 1936) for oxide glasses in the 1930s. Such radial distribution functions, however, describe only volume average positions of atoms that can be deduced from diffraction experiments and are not adequate for the understanding of the mechanism of inelastic deformation. It has been recognized for some time now that the "free volume," which refers to the fraction of matter having a lower atomic coordination than that in a reference material having a dense random packing of molecules or chains of molecules and the same composition, is of key importance for such purposes. It is in these free volume regions where mechanical coupling to the surroundings is weak that inelastic relaxations become possible by local atom rearrangements or molecular segment rotations without significantly affecting the surroundings. On the other hand, through the study of the kinetics of the linear viscoelastic behavior of glasses of all types under low stress, it has been established that such local relaxation processes are not monoenergetic but are characterized by a wide spectrum of activation energies that correspond to a wide gradation of free volume sites with different local coordination. For the purpose of conceptualization, it is useful to refer to reference structures of dense random-packed hard spheres such as the model of Bernal (1964) as an idealization for a "completely dense" simple glass. As is well known, the "Bernal glass" is made up of a characteristic distribution of five variously distorted polyhedral shapes ranging from tetrahedra to tetragonal dodecahedra that describe the sites of low atomic coordination of the dense reference glass. In real glasses that have been cooled at finite rates, the average atomic coordination will be lower than in the Bernal glass, giving rise to some free volume distributed over a continuous spectrum of sizes that will depend upon the cooling rate. Such a spectrum is likely to have a peak at somewhat above the size of a normal interstitial site in the corresponding ordered structure and a long but sharply decreasing tail of sites with larger volume. A very instructive visualization of this distribution can be obtained from a two-dimensional soap-bubble model of a glass that has been investigated by Argon and Kuo (1979) as an analog of a three-dimensional metallic glass. When properly constructed, according to the rules set down by Bragg and co-workers (Bragg and Nye, 1947; Bragg and Lomer, 1949), the soap-bubble glass is capable of giving quantitatively accurate information for a two-dimensional structure. Figure 1 shows a region of such a glassy bubble raft

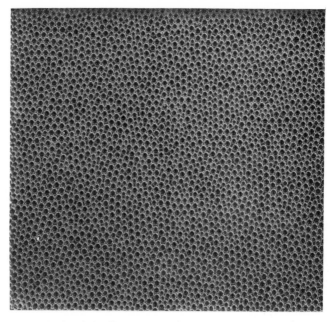

FIG. 1. Photograph of a glassy bubble raft made of two sizes of bubbles 1.38 and 1.08 mm in diameter in a number ratio of 1.32 (from Argon and Kuo (1979), courtesy of Elsevier).

composed of two different sizes of bubbles, while Fig. 2 gives the distribution of the excess interstital area determined by Argon and Kuo from such a raft. It is generally understood that at a temperature well below that of glass transition, T_g, in the undeformed state this free volume distribution of the structure is fixed. Furthermore, if the structure is aged below T_g, it is expected that after a relatively rapid stabilization further changes of the structure that require longer-range diffusion will be quite slow. It is clear that the above picture will be more satisfactory to describe a reasonably close-packed metallic glass, but that important differences will be present

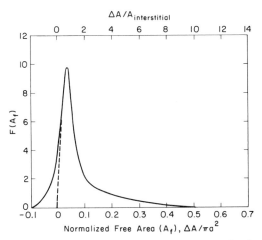

FIG. 2. Distribution of free area in the bubble raft sample shown in Fig. 1. ΔA represents the excess interstitial area over the reference area of an interstitial in a close-packed structure having a single average-size bubble (from Argon and Kuo (1979), courtesy of Elsevier).

in oxide glasses and glassy polymers. In the first case, strong covalent directional bonds will maintain a near-neighbor coordination of chemically bonded species such as silicon–oxygen tetrahedra, giving rise to a distorted network that will be broken only at the sites of network modifiers. Hence it can be expected that the free volume sites will chiefly be made up of the surroundings of these network modifying ions. In glassy polymers, on the other hand, the large free volume will derive from the impossibility of obtaining close packing of molecular segments of randomly oriented inter-penetrating polymer chains. In carbon chain polymers of relatively high chain flexibility, a random segment orientation is likely to be compatible with a modest free volume distribution. In stiff chain polymers, such as polyimides where total randomness in local chain segment orientation would result in very large interstitial holes and consequently large concen-trations of free volume, some local ordering between segments of mole-cules is probably inevitable. Such ordering is discernible from the results of deformation experiments of Argon and Bessonov (1977a,b), which shall be discussed in more detail in Section D. Outside of these quantitiative differ-ences in the description of free volume in these different glasses, we expect overriding qualitative similarities in its distribution and in its effects on the mode of inelastic deformation. Fundamentally then, it will be this free vol-ume and its distribution that differentiates glassy structures as weakly cou-pled mechanical systems of ions and molecules from crystalline substances that are always strongly coupled mechanical systems.

D. INELASTIC DEFORMATION IN THE ABSENCE OF STRUCTURE CHANGE

As discussed in the preceding section, at a temperature below T_g, in at least partially stabilized glass, we expect that the free volume distribution will change only very slowly with time so that it can be considered fixed for the purposes of discussing the mechanism of inelastic rearrangements. At temperatures very near to T_g and over very long periods of time, structure changes will of course persist and will strongly influence the mechanism of deformation. For a theory of plastic flow in glasses near T_g, we refer the reader to Robertson (1966). We shall discuss in this chapter only the deformations occurring in structures that are stable and fixed, i.e., *isostructural* deformation. Hence, we expect that the local inelastic rearrangements among atoms or molecular segments will be in the nature of local shear transformations in a quasi-fixed structure of quasi-fixed free volume distribution. Anticipating possibilities for strain localization, we are, however, prepared to encounter shear-induced alterations of free volume that may have catalyzing effects on further deformation. Normally, such local shear transformations can be thermally assisted under applied shear stresses, and their kinetics can be treated by the standard procedures of transition state theory [for a discussion of the application of transition state theory to crystal plasticity cf Kocks *et al.* (1975)].

According to the procedures of transition state theory of thermally activated phenomena, a flow unit such as a free volume site and its immediate surroundings is identified. Then the modes of internal rearrangement of the site that result in the production of shear strain at the distant boundaries are described by a set of normal mode parameters of the flow unit. The Helmholtz free energy contour of the rearrangement is constructed as a function of these normal mode parameters (or as a function of the increments of shear strain experienced at the distant boundaries to which the normal mode parameters relate) by conceiving a continuous set of reversible processes involving the alterations of the flow unit in equilibrium with the tractions exerted at the distant boundaries, and in equilibrium with all other relevant parameters describing the flow unit and its surroundings. In most instances, a single principal normal mode parameter of coupled interatomic motions governs the alteration of the flow unit and describes the *activation path* of the system over the saddle point of the potential contour. The free energy for activation of flow units under a given applied shear stress σ, less than the mechanical threshold stress $\hat{\tau}$ that can transform the flow unit at a terminal rate governed by inertia effects alone, can then be obtained by computing the deficiency in the free enthalpy* ΔG between the

---

*The term free enthalpy is preferred over the term Gibbs free energy, as in the usual discussions of thermodynamics, the latter is generally assumed to depend only on the isotropic

stable and unstable equilibrium positions of the system under the applied stress σ,

$$\Delta G = \Delta F - \Delta W. \tag{1}$$

In Eq. (1), ΔF, and ΔW are, respectively, the changes in Helmholtz free energy of the system and the deformation work done *on the system* at the distant boundaries by the applied tractions σ as the system transforms between its stable and unstable equilibrium positions. Finally, the thermal activation of the flow units is then governed by the availability of thermal fluctuations equaling or exceeding this required amount of ΔG. In the classical range where $kT > h\nu$ (ν is the normal mode frequency of the system in the activation coordinate), this availability is described by the Maxwell–Boltzman distribution and results in an externally measured rate process given by an Arrhenius expression.

Under a small applied stress $\sigma << \hat{\tau}$, at temperatures well below T_g, only a few and isolated free volume sites with the smallest free energies can be transformed in meaningful times of laboratory experiments while much of the background structure acts substantially as an elastic medium. This produces primarily delayed elasticity (anelasticity) through which the structure becomes mechanically "polarized" with only a minimum of accompanying irrecoverable deformation. Under substantially higher stresses, larger fractions of flow units with higher threshold stresses can be transformed into sheared states until at a certain well-defined stress the shearable flow units give contiguously sheared regions throughout the volume. Since the initial structure is a *random* packing of hard spheres, or tetrahedra in nonpolymeric glasses, the sheared structure inside these regions should also be *random* and statistically similar to the initial structure. Thus the original structure is repeatedly regenerated together with a capability for further shear deformation in the already sheared regions. Furthermore, contiguity at high stress relaxes the undeformed elastic background and results in loss of memory for the original state—making it possible for arbitrarily large shear deformations to take place. In fact, the only memory stored in the deformed structure is due to the continued mechanical polarization of the small fraction of large free volume flow units that have undergone shear transformations in the most recently regenerated structure during the very last phases of the deformation history prior to the removal of stress. In this description of inelastic deformation, the close-packed regions and the regions with the smallest free volume need not undergo any shear-

---

portion of the stress tensor. It is understood here that the free enthalpy is affected primarily by shear stresses but that pressure (or even individual elements of normal stress) can have secondary influences on the free enthalpy.

ing at all: they are merely trapped in the sea of deforming material around them and tumble along.

In long-chain polymer glasses, the process of deformation will require some special considerations. First, the modes of shear transformation of small regions will be subject to the constraint of preserving connectivity of individual molecular chains in the flow unit. This will usually mean that the fundamental mode of deformation will be rotation of individual segments on molecules in flexible chain polymers or, at most, cooperative and synchronized rotation of small bundles of adjoining molecule segments in some stiff chain polymers. Furthermore, upon continued shear of local regions, we expect to find the development of molecular orientation and a gradual change of the structure and its free volume distribution.

As briefly discussed above and recently by Spaepen (1977) and Argon (1979), we expect further that when relatively closely packed metallic glasses or polymeric glasses are sheared, transitory local dilatations may be produced. At low temperatures, where the diffusive decay of such transitory activation dilatations may be very slow, the sheared regions may retain a large "nonequilibrium" component of shear-induced excess free volume. Even at higher temperatures, however, where the shear-induced activation can decay reasonably rapidly, deformation at high rate may still result in a steady-state dynamic excess free volume. In metallic glasses where no important strain hardening or strain softening is likely and in polymeric glasses where orientation hardening is initially very weak, this dynamic excess free volume plays an important role in lowering the average deformation resistance $\hat{\tau}$ of the sheared regions and gives rise to strong shear localization. In the relatively open structures of inorganic glasses, dilatations induced by local shear transformations will either be very weak or, if present, will have very weak effects. Hence, we expect to find relatively strong strain localization effects in metallic and polymeric glasses leading to a high probability of shear band formation. In distinction, such effects should be nonexistent in inorganic glasses.

Based on this general description we consider next the mechanism of plastic flow in glasses in detail.

E. PLASTIC DEFORMATION IN GLASSES BY LOCAL SHEAR
   TRANSFORMATIONS

1. General Considerations

Argon and Kuo (1979) have studied the mechanism of plastic flow in a glassy soap-bubble raft designed to simulate an interbubble potential that closely resembles that in a metallic glass. When such soap-bubble rafts as shown in Fig. 1 were sheared and photographed, the modes of inelastic bubble rearrangement could be studied in detail. Such observations have revealed two limiting forms of rearrangement. In one type of rearrangement, occurring preferentially at a large free area (free volume) site, a complex and cooperative internal exchange occurs between bubbles in a relatively equiaxed region resulting in an effective shear transformation inside the region as depicted in Fig. 3a. In the other type of rerrangement, a

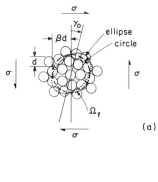

(a)

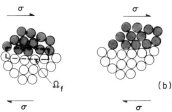

(b)

FIG. 3. (a) Diffuse shear transformation in a roughly equiaxed region; (b) shear translation in disk-shaped region between two quasi-close packed planes (from Argon (1979), courtesy of Pergamon Press).

sharp shear translation occurs as depicted in Fig. 3b between two relatively close-packed layers having particularly small free area (free volume). These rearrangements directly confirmed that the form of plastic deformation in glasses involves atom exchanges of a dipolar nature as imagined earlier by Orowan (1952), but that these exchanges are of a somewhat less local and more cooperative nature involving a larger number of atoms than previously recognized. It is clear that in both cases the local shear occurring either in an equiaxed region or in a flat prismatic region is in the nature of a shear transformation.

In extrapolating these rearrangements into a three-dimensional metallic glass, we imagine them to remain as shear transformations but to become of a rotationally symmetric form of a roughly spherical region and a flat,

penny-shaped region, respectively. We note that from energetic considerations alone the regions of shear transformation should all be flat and
penny-shaped, as this would store the smallest amount of elastic strain
energy in and around the sheared region (Eshelby, 1957). It is likely that
the choice of a spherical transformation region is dictated by its internal
constitution and by the limited modes of atom exchanges that it may offer.
By necessity therefore, if the elastic strain energy is to be of manageable
proportion, we expect that the transformation shear strain will be smaller
in the equiaxed regions than in the flat penny-shaped regions. This is
directly confirmed in the observations on the glassy bubble raft.

In the oxide glasses with their strong directional bonds, we expect that
the resistance to distortions of the network will favor shear transformations
in equiaxed regions but that more intense shears should be possible around
network defects at network modifiers or between such closely spaced network defects. In either case, it is inevitable that the transformations will
require some breakage and reestablishment of primary bonds as exchanges
occur between tetrahedra. It is clear that a direct extrapolation of the find-
ings in the bubble glass is less satisfactory for oxide glasses.

In glassy polymers the local process that gives rise to a shear transformation has been envisioned by many investigators (Robertson, 1966;
Argon, 1973a; Yannas and Lunn, 1975; Bowden and Raha, 1974) to involve
rotation of segments of chain molecules as depicted in Fig. 4.

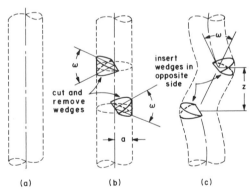

(a)   (b)   (c)

FIG. 4. Shear by molecular segment rotation in a polymer. The process is equivalent to the
production of a kink pair in a cylindrical region in a solid (from Argon (1973a), courtesy of
Taylor and Francis).

Thus, in every instance we expect that as the region transforms by a
thermal fluctuation under a shear stress σ between an elastically flexed
stable equilibrium state and an unstable equilibrium saddle-point configuration, on its way to a fully internally rearranged and sheared state, there

will be several contributions to the free enthalpy change ΔG of the system and its set of external tractions. The Helmholtz free energy will now have two components. First, the rearrangement of atoms in the region will change their potential energy ΔF as they undergo locally large exchanges. Second, the shear transformation to the saddle-point configuration will store elastic strain energy $\Delta \mathscr{E}$ both inside and outside the transformed region. Finally, the transformation will produce motions at the distant boundaries and will permit the applied shear traction σ to do an amount of work ΔW on the system, giving in total a free enthalpy change of

$$\Delta G = \Delta F + \Delta \mathscr{E} - \Delta W. \tag{2}$$

2. Plastic Flow in Metallic Glasses

Argon (1979) has evaluated the various contributions to the free enthalpy discussed in the preceding section for the two mechanisms of inelastic relaxation depicted in Fig. 3.

a. Deformation at High Temperature ($T < T_g$) and Low Stress ($\sigma \ll \hat{\tau}$). This favors shear transformations in equiaxed regions for which Argon has developed the following forms:

$$\Delta \mathscr{E} = [(7 - 5\nu)/30(1 - \nu)]\mu\gamma_0^2\Omega_f, \tag{3a}$$

$$\Delta F = \hat{\tau}\gamma_0\Omega_f \qquad (\sigma \ll \hat{\tau}), \tag{3b}$$

$$\Delta W = \sigma\gamma_0\Omega_f, \tag{3c}$$

where Ω_f is the roughly spherical volume of the transformed region of radius R, γ_0 the transformation shear strain, μ the shear modulus, ν the Poisson ratio, $\hat{\tau}$ the maximum gradient of the interatomic potential energy with respect to shear distortion (the athermal threshold plastic resistance at the given temperature), and σ the applied shear stress.

When the possibility of reversal of deformation is also taken into account in the transformed region by a reverse thermal fluctuation doing work against the applied shear stress, the total strain rate $\dot{\gamma}$ is found to be of the form

$$\dot{\gamma} = \alpha\gamma_0\nu_G\exp\left(-\frac{\Delta\mathscr{E} + \gamma_0\hat{\tau}\Omega_f}{kT}\right)\sinh\left(\frac{\sigma\gamma_0\Omega_f}{kT}\right), \tag{4}$$

where ν_G is the normal mode frequency $[0(10^{12})\ \text{sec}^{-1}]$ of the transforming complex along the activation path, and $\alpha\ [0(1)]$ incorporates numerical constants and the steady-state volume fraction of flow units contributing to plastic flow. The stress exponent m of the strain rate in the phenomenologically useful power function approximation of Eq. (4) is given by

$$m \equiv \left(\frac{\partial \ln \dot{\gamma}}{\partial \ln \sigma}\right)_T = \left(\frac{\sigma}{\mu}\right)\left(\frac{\mu\gamma_0\Omega_f}{kT}\right)\coth\left(\frac{\sigma}{\mu}\frac{\mu\gamma_0\Omega_f}{kT}\right), \tag{5}$$

where we note that as $\sigma/\hat{\tau}$ goes to zero in the limit, $m \to 1$ as expected. We obtain the temperature dependence of the flow stress by inverting Eq. (4) and get

$$\sigma = \left(\frac{kT}{\gamma_0 \Omega_f}\right) \sinh^{-1}\left[\left(\frac{\dot{\gamma}}{\alpha \gamma_0 \nu_G}\right) \exp\left(\frac{\Delta \mathscr{E} + \gamma_0 \hat{\tau} \Omega_f}{kT}\right)\right]. \tag{6}$$

b. Deformation at Low Temperature ($T \ll T_g$) and High Stress ($\sigma < \hat{\tau}$). This favors shear transformations in narrow penny-shaped regions for which obtaining ΔG^*, the activation free enthalpy, requires detailed consideration of the parameters governing the size and internal shear translation of the activation configuration of Fig. 3b at the saddle point. Argon (1979) has demonstrated that the detailed and complex dependence of the activation free enthalpy on stress can be conveniently summarized in a simple phenomenological form given by

$$\Delta G^* = 4.56 \hat{\tau} \Omega_f \left(1 - \frac{\sigma}{\hat{\tau}}\right)^2, \tag{7}$$

where Ω_f, the volume of the narrow sheared configuration, is to be taken as $a\pi R^2$ with the thickness a and the diameter of the sheared region $2R$ being of the order of one and five atomic diameters d, respectively. As in the case of the diffuse shear transformation discussed, in the sharp shear translation, the resulting shear strain rate can be given as

$$\dot{\gamma} = \alpha \gamma_0 \nu_G \exp\left(-\frac{\Delta G^*(\sigma/\hat{\tau})}{kT}\right). \tag{8}$$

The stress exponent m in this case acquires the form

$$m = -\frac{\sigma}{kT}\left(\frac{\partial \Delta G^*}{\partial \sigma}\right) = 4.56\left(\frac{\hat{\tau}\Omega_f}{kT}\right)\frac{\sigma}{\hat{\tau}}\left(1 - \frac{\sigma}{\hat{\tau}}\right). \tag{9}$$

Finally, the temperature dependence of the flow stress at constant strain rate can be obtained from Eqs. (7) and (8) as

$$\frac{\sigma}{\hat{\tau}} = 1 - \left[0.22\left(\frac{kT}{\hat{\tau}\Omega_f}\right)\ln\left(\frac{\alpha\gamma_0\nu_G}{\dot{\gamma}}\right)\right]^{1/2}. \tag{10}$$

Although the plastic behavior for many metallic glasses at low temperature (Davis, 1976, 1978; Pampillo, 1975) has been reported, reliable data are sparse because deformation localization and premature fracture seriously affect the results. Some properly performed experiments do, however, exist in which yield stresses were measured either in compression, where fracture does not terminate the experiment (Pampillo and Chen, 1974), or at temperatures near T_g, where no deformation localization occurs (Megusar *et al.*, 1979). These results have been compared by Argon (1979) with the theoretical results given above and have been found to be in good agreement, as shown in Figs. 5 and 6. There are currently no reliable measure-

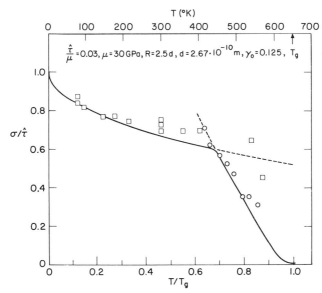

FIG. 5. Dependence of flow stress on temperature computed from Eqs. (6) and (10), and compared with experimental results of Pampillo and Chen (1974) and Megusar *et al.* (1979). ○, $Pd_{80}Si_{20}$ ($\hat{\tau} = 0.90$ GN/m²) (Megusar *et al.*, 1979); □, $Pd_{77.5}Cu_6Si_{16.5}$ ($\hat{\tau} = 1.03$ GN/m²) (Pampillo and Chen, 1974) (from Argon (1979), courtesy of Pergamon Press).

ments of the strain rate sensitivity of the flow stress in the low-temperature regime. All indications, however, broadly support the theoretical prediction that this effect will be very small, i.e., that the stress exponents m will be very high. To obtain the observed match shown in Figs. 5 and 6, the following physical constants were used: shear modulus $\mu = 30$ GPa, Poisson's ratio $\nu = 0.3$, average atomic radius $d = 2.67 \times 10^{-10}$ m, $\alpha\gamma_0\nu_G = 10^{11}$ sec⁻¹, ideal elastic shear strain at yield $\hat{\tau}/\mu = 0.03$. This has required the choice of the transformation shear strain $\gamma_0 = 0.125$, and the radius of the transformed region $R = 2.5d$. These last two values of the "fitting" constants are inspired by the bubble model and are considered to be very satisfactory (Argon, 1979).

3. Plastic Flow in Oxide Glasses

It is generally accepted that unmodified oxide glasses can be considered as distorted space networks, while modified oxide glasses contain network defects at sites of network-modifying ions. This structural description and

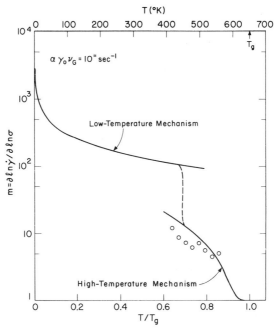

FIG. 6. Temperature dependence of the stress exponent m at plastic flow computed from Eqs. (4), (5), (8), and (9) and compared with experimental results of Megusar *et al.* (1979) (○, Pd$_{80}$Si$_{20}$). (From Argon (1979), courtesy of Pergamon Press).

the generally homogeneous nature of flow in them suggests that plastic deformation, when at all possible, should be primarily by means of the high-temperature mechanism of shear transformations in roughly spherical regions introduced above—modified appropriately to be valid over the entire stress range. Based on this mechanism, a simple development gives the stress dependence of the activation free enthalpy for the local shear transformation as (see Appendix)

$$\Delta G^* = [\hat{\tau}\Omega_f\gamma_0/4(1 - B)][1 - (\sigma/\hat{\tau})]^2, \tag{11}$$

where

$$B = [(7 - 5\nu)/30(1 - \nu)](\mu\gamma_0/\hat{\tau}) \tag{12}$$

is the scale factor of the modified elastic energy stored around the transformed region [see Eq. (3a)]. As before, we expect the shear strain rate to be given by an Arrhenius expression,

$$\dot{\gamma} = \alpha\gamma_0\nu_G \exp\left(-\frac{\Delta G^*(\sigma/\hat{\tau})}{kT}\right), \quad \sigma/\hat{\tau} > 0. \tag{8a}$$

The stress exponent m of the strain rate and the temperature dependence of the flow stress at constant strain rate are then obtainable as before and are

$$m = -\frac{\sigma}{kT}\left(\frac{\partial \Delta G^*}{\partial \sigma}\right)_T = \frac{\hat{\tau}\Omega_f\gamma_0}{(2kT(1-B)}\frac{\sigma}{\hat{\tau}}\left(1 - \frac{\sigma}{\hat{\tau}}\right), \tag{13}$$

$$\frac{\sigma}{\hat{\tau}} = 1 - \left[\frac{4(1-B)kT}{\hat{\tau}_f\Omega_0}\ln\left(\frac{\alpha\gamma_0\nu_G}{\dot{\gamma}}\right)\right]^{1/2} \tag{14}$$

Primarily because of the inherent brittleness of oxide glasses, reliable plastic flow data are only available from microhardness indentations where quite often the very small size of the indentation makes it possible to avoid surface flaws and to obtain fracture-free plastic indentations. Even though such experiments are not difficult, they have been performed systematically only by Marsh (1964a,b) on a soda-lime silica glass (window glass), and a high modulus "E" glass. His results are compared with the theoretical model in Figs. 7 and 8 by the solid curve and are found to be in fair

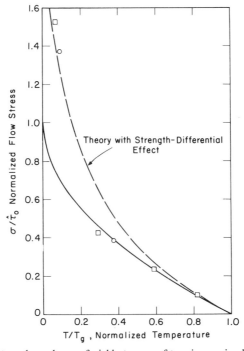

FIG. 7. Temperature dependence of yield stresses of two inorganic glasses obtained from hardness indentation measurements of Marsh (1964a,b), and compared with theoretical model given by Eqs. (14) and (17). ○, Soda glass; □, E glass.

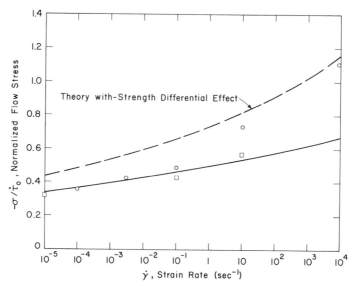

FIG. 8. Strain rate dependence of yield stresses of two inorganic glasses obtained from hardness measurements of Marsh (1964a,b), and compared with theoretical model given by Eqs. (14) and (17). ○, Soda glass; □, E glass.

agreement if the following choices are made for the physical parameters in Eqs. (11)–(14): $\gamma_0 = 0.125$ (same as for metallic glasses), $\hat{\tau}/\mu = 1/2\pi$ (as predicted from a sinusoidal shear resistance), $\Omega_f = 1.496 \times 10^{-27}$ m^3, and $\nu_G = 1.575 \times 10^8$ sec^{-1} obtained from a match with the strain rate dependence of the flow stress and a choice of $\alpha = 0.5$. The very low temperature and very high strain rate results of Marsh's measurements are, however, not accounted for by the simple theory presented above. The explanation of this systematic departure lies in the well-known strength-differential (SD) effect that results from the increase of the ideal shear strength of a solid with increasing pressure. According to Kelly $et\ al.$ (1967), this effect in ionic solids is roughly of a magnitude of

$$\hat{\tau}_i = \hat{\tau}_{i0} + 0.5p, \tag{15}$$

where $\hat{\tau}_{i0}$ is the ideal shear strength of the solid measured in a shear experiment, and p the superimposed pressure of any other stress state. Marsh (1964a) has shown that in strong solids, where large elastic distortions accompany plastic flow under an indenter, the plastic constraint effect is not as large as in nearly "rigid" plastic solids that yield under much lower stresses in proportion to their moduli. Based on experiments and approximate computations of Marsh, the average superimposed pressure p under an indenter during plastic flow in a strong solid can be given for our purposes approximately as

$$p \simeq 12.84\hat{\tau} \ \ln\left(1 + 0.105\sigma/\hat{\tau}\right), \tag{16}$$

where $\hat{\tau}$ is the shear deformation resistance of the solid under zero pressure and σ the flow stress in shear in the indenter flow field calculated on the basis of a von Mises yield theory. Interpreting $\hat{\tau}_{i0} = \hat{\tau}$ as has been the practice in the preceding section, the strength-differential effect can be incorporated into Eqs. (8a), (13), and (14) by replacing $\sigma/\hat{\tau}$ with

$$\sigma/\hat{\tau} \rightarrow (\sigma/\hat{\tau}_0) \ [1 + 6.42 \ \ln\left(1 + 0.105\sigma/\hat{\tau}_0\right)], \tag{17}$$

where $\hat{\tau}_0 = \hat{\tau}_{i0}$. Solution of Eqs. (8a) and (14) with this modification gives the broken curves in Figs. 7 and 8, which now adequately account for the low-temperature and high-strain-rate measurements. It is clear that, for better agreement along the entire range, the pressure distribution in an elastic–plastic indentation field of a strong solid needs to be known better together with a more accurate representation of the effect of pressure on the shear resistance in oxide glasses.

The observations of plastic flow in oxide glasses under suitable conditions, particularly under indenters, is not limited to the experiments of Marsh alone. Similar observations were made independently by Douglas (1958), Peter (1964), Hillig (1962), and others. Bridgman and Simon (1953) have reported large plastic strains for thin disks compressed between anvils under very high pressure. These strains were accompanied by substantial magnitudes of volumetric compaction indicating that the open structure of oxide glasses permits large semipermanent inelastic compaction strains. These, however, are recoverable by means of a high-temperature anneal while plastic shear strains are not. Ernsberger (1968), who has made careful measurements of birefringence of microindentation regions in glasses, has found that substantial volumetric compaction occurs there and has therefore counseled caution in attributing all indentation strains to plastic shear flow. It can be readily shown, however, that even after prolonged high-temperature anneal, about half the indentation remains and can therefore be attributed to plastic shear flow (Argon, 1969). Other observations of plastic flow in glasses include crack-free grooves produced by scribers on glass surfaces by Peyches (1952) and Marsh (1964a); crack-free, helicoidal microscopic "machining chips" of glass, obtained by lightly dragging a very sharp point of a tungsten carbide tool on glass surfaces, seen by McClintock and Argon (1966); rapid increases of fluidity in glasses with increasing stresses, at stresses around 1% of the shear modulus, measured by Li and Uhlmann (1970a,b); and finally, anomalously high-fracture toughnesses in glasses reported by Kerkhof and Richter (1969). While the last cited example is contrary to basic expectations and is subject to some doubt, all the other examples of plastic flow are incontestable and indicate that plastic flow can be enforced in oxide glasses under suitable conditions. It is worthwhile to reemphasize, however, that plasticity in oxide glasses even under the most favorable conditions is very limited and of little technological importance. Therefore, oxide glasses can be taken as intrinsically brittle solids that will fracture before any significant plastic flow can occur under states of stress not containing a very high superimposed pressure component.

4. Plastic Flow in Glassy Polymers

As discussed briefly above, plastic flow in a glassy polymer of long-chain molecules is accomplished not by severing molecules but by producing thermally acitvated rotations of short segments of molecules. This incremental process proceeding in time produces eventual alignment of molecules parallel to the imposed principal strain direction. Hence, unlike in metallic or oxide glasses, in polymeric glasses plastic flow produces strain hardening and development of strong anisotropy. When the plastically hardened and oriented polymer is brought to the glass transition temperature where intermolecular resistances to deformation are effectively removed, it will usually return to its initial shape of maximum configurational entropy of random segment orientations. This complete recoverability of plastic strain is shown in Fig. 9 for polymethylmethacrylate (PMMA).

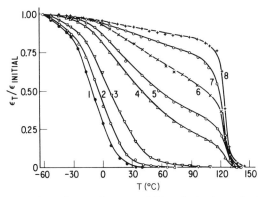

FIG. 9. Temperature dependence of recoverability of plastic strain in PMMA samples strained to: (1) 0.06; (2) 0.077; (3) 0.115; (4) 0.179; (5) 0.21; (6) 0.30; (7) 0.49; and (8) 0.668 (after Arzhakov et al., as quoted by Argon and Bessonov (1977a,b), courtesy of Polymer Eng. Sci.).

It indicates that strain hardening in polymeric glasses is far from isotropic and is of a *kinematic* type.

By following the procedure outlined in Sections II.D and II.E.1, and by using a specific solution of Li and Gilman (1970), evaluating the changes of intermolecular strain energy in and around a molecular segment of average radius a and of length z undergoing a rotation ω in opposition to the surrounding molecules, Argon (1973a) has calculated the free enthalpy change ΔG^* associated with the saddle-point configuration and has found it to be

$$\Delta G^* = \frac{3\pi\mu\omega^2 a^3}{16(1-\nu)}\left[1 - 8.5(1-\nu)^{5/6}\left(\frac{\sigma}{\mu}\right)^{5/6}\right], \tag{18}$$

where σ is the applied shear stress, μ the shear modulus, and ν the Poisson ratio. Furthermore, this saddle-point configuration prescribes the most suitable length of segment z^* of a molecule undergoing a rotation ω fixed by the chemistry of the molecule as

$$z^*/a = \{[45/8(1-\nu)]\,(\mu/\sigma)\}^{1/6}. \tag{19}$$

As in the previous cases, the strain rate is given by an Arrhenius equation of the type of Eq. (8) with a preexponential factor $\dot{\gamma}_0$ having substantially the same composition as in the other glasses discussed. This permits immediately the description of the stress exponent m of the strain rate and the temperature dependence of the flow stress as

$$m = \frac{1.33\pi}{(1-\nu)^{1/6}}\left(\frac{\mu\omega^2 a^3}{kT}\right)\left(\frac{\sigma}{\mu}\right)^{5/6}, \tag{20}$$

$$\frac{\sigma}{\hat{\tau}} = 1 - \frac{16(1-\nu)kT}{3\pi\mu\omega^2 a^3}\ln\left(\frac{\dot{\gamma}_0}{\dot{\gamma}}\right), \tag{21}$$

$$\hat{\tau} = \frac{0.077}{(1-\nu)}\mu, \tag{22}$$

where $\hat{\tau}$ as before is the athermal flow resistance at the given temperature. Argon and Bessonov (1977a,b) have examined the yield behavior of a large number of glassy polymers, including not only the familiar carbon chain polymers of polymethylmethacrylate (PMMA), polystyrene (PS), polycarbonate (PC), polyethylene terephthalate (PET), and aromatic polymers such as polyphenylene oxide (PPO), but also a group of polyimides such as resorcinol (R–R), hydroquinone (H–H), oxydiphenyl (DFO), and pyromellitic acid (PM, Kapton), and have compared the behavior of these polymers with the predictions of the theory. The chemical structure of these polymers, their glass transition temperatures T_g, their average molecular radii a_0, and the lengths $\bar{l}$ of the stiff units (segments) between natural hinges on the molecules are given in Table I (Argon and Bessonov, 1977a,b). In five instances, the lengths $\bar{l}$ are exactly equal to the lengths of the monomer units. It is clear from Table I that in comparison with the simple carbon-chain (PS, PMMA) polymers, the aromatic polymers (PPO and polyimides) represent a separate group of increased natural hinge length. The polymers of intermediate chemical structure, PC and PET, have intermediate char-

TABLE I

Chemical Structure and Physical Properties of Some Glassy Polymers

No.	Polymer	Monomer unit	T_g (°C)	a_0 (10^{-10} m)	$\bar{l}$ (10^{-10} m)	$\bar{l}/a_0$
1	Kapton (PM)	*(chemical structure)*	>350[a]	2.82	18.0[b]	6.4
2	DFO	*(chemical structure)*	270	2.94	11.5[b]	3.9
3	H–H	*(chemical structure)*	230	—	8.6	~3.0
4	R–R	*(chemical structure)*	200	—	8.1	~2.8
5	PPO	*(chemical structure)*	170	3.36	5.6[b]	1.7
6	PC	*(chemical structure)*	150	3.09	2.8	0.9
7	PET	*(chemical structure)*	65	2.84	2.15	0.8
8	PS	*(chemical structure)*	100	4.63	1.55[b]	0.35
		(chemical structure)	110	2.85	1.55[b]	0.55

[a] No true glass transition.
[b] Exactly equal to the length of stiff units on the chain.

100

acteristics. In comparison to $\bar{l}$, there is no great disparity between the average molecular radii a_0 in all polymers shown in Table I, with the exception of PS and PPO, which have relatively bulky side groups.

To facilitate the comparison of the theory with experimental results, Eq. (21) can be cast in a simpler form

$$(\sigma/\mu)^{5/6} = A - B(T/\mu), \tag{23}$$

where the constants A and B are

$$A = [0.077/(1 - \nu)]^{5/6}, \tag{24}$$
$$B = A\{[16(1 - \nu)k/3\pi\omega^2 a^3]\,ln(\dot{\gamma}_0/\dot{\gamma})\}. \tag{25}$$

If the plots of the experimental data of $(\sigma/\mu)^{5/6}$ as functions of T/μ give straight lines having a common intercept at absolute zero temperature, the different slopes of the lines will reflect the differences in the molecular parameters. These can be obtained as

$$\omega^2 a^3 = (A/B)[16(1 - \nu)k/3\pi]\,ln(\dot{\gamma}_0/\dot{\gamma}), \tag{26}$$
$$(z/a)^*_0 K = [45/8(1 - \nu)]^{1/6}\,A^{-1/6}. \tag{27}$$

The normalized plastic shear resistance functions $(\sigma/\mu)^{5/6}$ for all polymers are plotted as functions of T/μ in Fig. 10. As can be seen in all instances, the data lie along straight lines in accordance with the theory. Data points very near the respective T_g deviated sharply from the slanted lines in the direction parallel to the T/μ axis. This is the so-called "leathery" region, where appreciable free volume changes occur with increasing

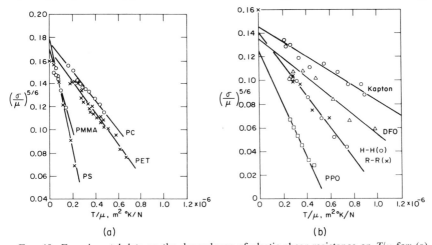

(a)   (b)

FIG. 10. Experimental data on the dependence of plastic shear resistance on T/μ for: (a) carbon chain polymers, (b) polyphenylene oxide and polyimides. The star on the ordinate axes indicates theoretically predicted intercepts for $\nu = 0.3$ (from Argon and Bessonov (1977a,b), courtesy of Taylor and Francis).

temperature, and falls outside the scope of the theory for deformation in a structure of temperature-independent free volume. Such points were therefore not plotted in Fig. 10.

According to the theory, all polymers should have a common intercept of $(\sigma/\mu)^{5/6} = A = 0.159$ (for $\nu = 0.3$) at $T = 0$ K, which has the meaning that at 0 K all polymers asymptotically reach their ideal shear resistance that is expected to be a constant of the glassy structure. The actual experimental lines, however, form two groups: the group of carbon-chain polymers of PS, PMMA, PET, and PC have common intercept of $A = 0.174$, while the group of true aromatic and heteroaromatic polyimides have an intercept around $A = 0.140$. This difference is most likely of geometrical origin related to the different molecular cross sections of these polymers. The Van der Waals cross sections of monomer units in the polyimides are of elliptical shape with axis ratios of 2:1 or larger.

As expected from Eq. (25), the most prominent differences between the data for different polymers are in the slopes of the lines in Fig. 10 that reflect directly the differences in the molecular structure of the polymers. The slopes are the same only for H–H and R–R, which have very similar chemical structures. The model characteristics of each polymer computed from the slopes of Fig. 10 according to Eqs. (26) and (27) are shown in Table II (Argon and Bessonov, 1977a,b). In these calculations, the following values were taken: $\nu = 0.3$, $\dot{\gamma}_0 = 3 \times 10^{13}$ sec^{-1} (Argon, 1973a), $\dot{\gamma} = 9 \times 10^{-3}$ sec^{-1}, which is the actual mean shear strain rate used in the experiments, and the mean valence angle $\omega = 2.0$ ($\sim 115°$) for all the molecules considered ($\sim 110°$ for the hydrocarbon polymers, and $\sim 120°$ for the oxy-aromatic polymers). Table II shows that the calculated radii a and the critical lengths z^* have "normal" molecular dimensions and change systematically between different polymers. The direct comparison of these experimentally obtained model parameters with the actual corresponding molecular dimensions a_0 and $\bar{l}$ gives the most significant insight. The ratios of $z^*/\bar{l}$ give the average number of molecular segments in the activated configuration. It is expected that this number indicates that z^* is a integer multiple of molecular segment lengths between natural hinges. The dimensions of z^* are coupled with the ratio of $(a/a_0)^2$ that gives the mean number of neighboring molecules acting collectively as a microbundle in the thermally activated local rearrangement. For reasons of thermodynamics, the ratio of z^*/a must remain at the value given by Eq. (19). The values of z^*/a at other temperatures will not differ significantly from the values of 0 K given in Table II. The computed ratios of $z^*/\bar{l}$ and $(a/a_0)^2$ are plotted in Fig. 11 as a function of the natural hinge-to-hinge distance $\bar{l}$ in the polymers, showing a clear trend in the results. Figure 11a shows that the ratio $z^*/\bar{l}$ goes asymptotically to unity as $\bar{l}$ becomes large, indicating that in such

TABLE II

MEASURED AND COMPUTED EXPERIMENTAL PARAMETERS ON THE PLASTIC FLOW OF GLASSY POLYMERS

No.	Polymer	Experimental A	B (MN/m²K)	$\omega^2 a^3$ (10^{-30} m³)	Calculated from experiments and structural data $(z/a)^*$ (0 K)	a (10^{-10} m)	z^* (10^{-10} m)	a^2/a_0^2	$z^*/\bar{l}$	References for experimental data on $\tau(T)$	$\mu(T)$
1	Kapton	0.145	0.060	1650	2.09	7.45	15.6	7.00	0.87	a	a
2	DFO	0.135	0.080	1240	2.12	6.77	14.4	5.27	1.25	a	a
3,4	R–R								1.48		
5	H–H	0.140	0.133	745	2.10	5.71	12.0	~3.60	1.40	a	a
	PPO	0.125	0.223	444	2.15	4.81	10.3	2.06	1.84	b	c
6	PC	0.176	0.129	768	2.01	5.77	11.6	3.46	4.14	d	d
7	PET	0.169	0.138	718	2.03	5.64	11.4	3.92	5.30	e	f
8	PS	0.180	0.480	206	2.00	3.72	7.44	0.65	4.80	g	h
9	PMMA	0.171	0.364	272	2.02	4.08	8.24	0.82	5.32	i	i

[a] Argon and Bessonov (1977a,b).
[b] Robertson (1968).
[c] Hay (1969).
[d] Bauwens-Crowet et al. (1969, 1972).
[e] Ward and Foot (1973).
[f] Illers and Breuer (1963).
[g] Argon et al. (1968).
[h] Rudd and Gurnee (1957).
[i] Bowden and Henshall (1973).

103

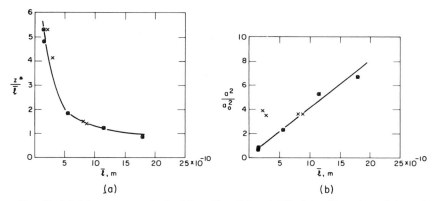

FIG. 11. Model parameters obtained from Eqs. (25) and (27) giving dimensions of the activated complex, as a function of the spacing $\bar{l}$ of natural "hinges" on the molecules: (a) $z^*/\bar{l}$, (b) $(a/a_0)^2$. Circles show cases where $\bar{l}$ is exactly equal to the known lengths of stiff units on the molecular chain (from Argon and Bessonov (1977a,b), courtesy of Taylor and Francis).

polymers with large $\bar{l}$ the long dimension of the activated complex is governed by the segment length between natural hinges, and that $z^*/\bar{l}$ can be no smaller than unity. The results in Table II and Fig. 11b show that in many instances $(a/a_0)^2$ is larger than unity, necessitated by the constraint of Eq. (19), indicating collective action of molecular segments in bundles. Note that in PS and PMMA the experiments give that there is only one molecule per bundle, in agreement with the recent extensive neutron scattering and x-ray studies that show no significant local order in typical carbon-chain polymers (Haward, 1973). In the case of other polymers where cooperative motions of larger numbers of neighboring molecules are calculated from the experimental results, the presence of some local ordering is suggested particularly for the aromatic and heterocycle polymers. This is consistent with the similarity of the chemical structure of the aromatic polymers and that of liquid crystals where aggregation of the molecules is the basis of their unique properties.

Figure 11 shows systematic variations of $z^*/\bar{l}$ and $(a/a_0)^2$ with $\bar{l}$ in all the polymers tested. This is not a mere artifact of the constraint on the activated configuration given by Eq. (19), but appears in the systematic trend in the computed $\omega^2 a^3$ with $\bar{l}$ shown in Tables I and II. Although there is no present mechanistic molecular understanding of this dependence, it is clear that for the case of the aromatic polymers the increased $\bar{l}$ is obtained without a significant change in chemical structure. It is suspected that polymers with large $\bar{l}$ are likely to have structures with molecular aggregation akin to liquid crystals, showing true nonlocal deformation modes involving generation and motion of dislocations already known to be of importance in the deformation of crystalline polymers.

In concluding the discussion of experimental results, it can be stated that the current experiments on aromatic polymers further support the notion that plastic flow in glassy polymers is of a highly local nature, but tends to become less so as the natural hinge spacing on molecules increases. Furthermore, the existing evidence suggests strongly that observations on shear localization effects in glassy polymers reported first by Whitney (1963) result from increased mechanical coupling between local molecular events. We shall return to this subject in Section II.G.

F. YIELDING UNDER A GENERAL STATE OF STRESS

Although pressure can influence the plastic shear resistance significantly, it cannot produce plastic flow on the large scale. As in the case of crystalline materials, large-scale plastic flow in glassy polymers requires a critical state of deviation of the stress tensor from a state of pure pressure. Hence, it is natural to expect that large-scale yielding in glassy polymers obeys a deviatoric stress criterion initially stated by von Mises (1913) where yielding occurs when the deviatoric stress s (root-mean-square shear stress) equals the plastic shear resistance $\sigma(T, \dot{\gamma}, p)$ for the given pressure component of the stress tensor, i.e.,

$$s \equiv \{ \tfrac{1}{6}[(\sigma_{22} - \sigma_{33})^2 + (\sigma_{33} - \sigma_{11})^2 + (\sigma_{11} - \sigma_{22})^2] + \sigma_{23}^2 + \sigma_{13}^2 + \sigma_{12}^2 \}^{1/2} = \sigma(T, \dot{\gamma}, p), \quad (28)$$

where $\sigma_{11}, \ldots, \sigma_{12}$ are the various components of the multiaxial stress state. Under biaxial stress this criterion leads to an ellipse, the center of which is translated into the third quadrant, along the line of pure pressure. Figure 12 shows biaxial yield data assembled by Raghava *et al.* (1973) for a number of polymers that confirm this expectation.

There have been a large number of other phenomenological yield criteria that have been advanced, particularly to explain the orientation of shear localization zones. Such criteria have often been deficient in adequately recognizing first, that large elastic distortions accompany plastic flow, which can produce significant apparent rigid body rotations of bands, and second, that formation of bands is a process of development of shear localization and is preceded by significant amounts of homogeneous plastic flow.

G. LOCALIZATION OF DEFORMATION INTO SHEAR BANDS

Except in network-forming oxide glasses and in highly cross-linked glassy polymers, the first increments of plastic strain can reduce the resistance of the material to subsequent deformation. Such strain softening is widely encountered in metallic and polymeric glasses at low temperatures and high strain rates. Two examples of such shear banding in a typical

A. S. ARGON

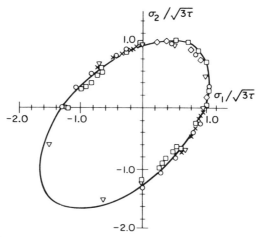

FIG. 12. Experimental results for biaxial yielding of several polymers ○, Raghava *et al.* (1973), (PVC); □, Raghava *et al.* (1973), (PC); ∇, Whitney and Andrews (1967), (PS); x, Bauwen (1970), (PVC); ◇, Sternstein and Ongchin (1969), (PMMA). (From Raghava *et al.* (1973), courtesy of Chapman and Hall).

metallic glass and a typical glassy polymer are given in Fig. 13. The origin of such shear localization lies in the alterations of local coordination between atoms and molecules. Argon (1979) has shown that the process in metallic glasses can be rationalized on the assumption that at low temper-

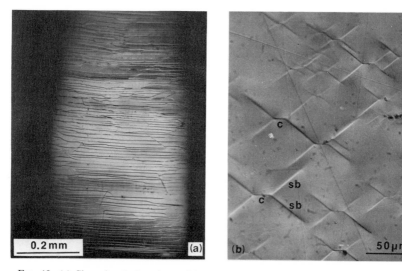

FIG. 13. (a) Shear bands in a bent ribbon of Metglas 2826B ($Fe_{29}Ni_{49}P_{14}B_6Si_2$); (b) shear bands *c* (and short crazes *sb*) in PET—tension direction is vertical.

atures and high strain rates, where local dilatations that may be produced during the unit shear processes cannot be readily diffusively dispersed, a steady-state reduced coordination, or excess free volume, can be retained in the sheared glass. Similar processes would be expected to hold in glassy polymers as well. If the principal effect of such excess free volume is a reduction of the terminal deformation resistance $\hat{\tau}$ of the glass according to some law similar to Eq. (15) prescribing rhe reduction in response to a negative pressure, then the rate of shear localization with additional increments of total shear strain are readily calculable by elementary concepts of compatibility and rules of mixing (Bowden and Raha, 1970; Argon, 1973b). The details of such simple computations of shear strain localization into bands have been described adequately elsewhere (Argon and Bessonov, 1977a,b; Argon, 1973b, 1979) and will not be reproduced here. In Fig. 14 we present a computed plot for shear localization intended for a typical metallic glass at low temperature, initiated by a 20% strain rate perturbation that may occur from surface irregularities (Argon, 1979). The figure shows

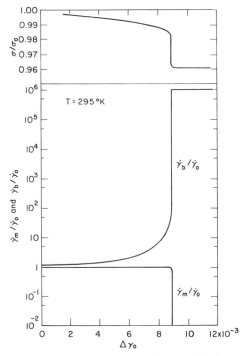

FIG. 14. A computed shear localization process in a $Pd_{80}Si_{20}$ glass at 295 K showing the concentration of shear strain rate into a narrow band while deformation rate outside the band ceases sharply (from Argon (1979), courtesy of Pergamon Press).

that if such concentration sites are few, very large shear strain rate concentrations can occur in narrow bands at the expense of the matrix in which deformation decelerates under a decreasing overall traction—all in a very small overall plastic shear strain increment. Cinematographic observations by Neuhäuser (1978) of the formation of shear bands in metallic glasses are largely in agreement with these predictions (Argon, 1979). The processes of shear localization in glassy polymers where molecular orientation produces eventual strain hardening are similar to those in metallic glasses but are free of the potentially runaway conditions that metallic glasses exhibit leading to complete shearing off (Megusar *et al.*, 1979).

III. Fracture in Glasses under Monotonic Loading

A. THE FRACTURE INSTABILITY

The role of cracks and flaws in brittle fracture under monotonic loading has been well appreciated since the pioneering papers of Griffith (Griffith, 1920, 1924). The modifications that are necessary to Griffith's theory to understand fracture in more or less plastically deformable solids have been subjects of intense interest to ever-widening groups of researchers. The basic directions to these developments were given by Orowan (1949) and by Irwin (1948), who in particular founded the branch of study of fracture mechanics that concerns itself with the precise statement of the condition of the fracture instability in structural components. Discussion of this very extensive subject will be outside the scope of this chapter. We shall find it sufficient to note that, depending upon the level of inelastic deformation that is necessary to propagate a crack in a solid, the condition of the fracture instability can be stated as a critical stress intensity factor K_I, a critical crack opening displacement δ_{c0} defined at the root of the crack, and a critical energy release rate G_I. These terms, defined for tension in relation to the macroscopic parameters under the control of the experimenter, are given for a plane strain setting as follows:

$$K_I = \sigma(\pi c)^{1/2} F(c/w), \qquad (29)$$
$$\delta_{c0} = \alpha K_I^2 / YE, \qquad (30)$$
$$G_I = -\partial V/\partial c = [(1 - v)^2/E] K_I^2. \qquad (31)$$

In Eqs. (29)–(31), σ is the applied tensile stress, c the half crack length, $F(c/w)$ a function of the specimen width w ($F \to 1$ for $c/w \to 0$), Y the yield strength in tension in a nonstrain hardening idealization, E Young's modulus, v Poisson's ratio, α a constant of order unity, and V the potential energy of the system of sample and its tractions. The functions $F(c/w)$ have been calculated for a large number of shapes in which the crack length is of finite proportions with respect to the width w and are readily available (Paris and Sih, 1965; Tada et al., 1973; Rooke and Cartwright, 1976). In glasses below T_g, where the inelastic deformations at the time of fracture are confined to the surroundings of the tip of the crack (small-scale yielding), the three alternative forcing functions given above are of equal utility. At temperatures very near T_g, where the glasses can become more compliant and tough so that the inelastic deformation zone spreads out over a large portion of the sample before the crack begins to propagate, different forcing functions based on nonlinear constitutive behavior become necessary. Since this falls outside our range of interest, however, we shall not expand on this topic further.

The subject of interest to us will be the mechanisms that govern the critical levels of these "forcing functions" for fracture in the different glasses.

The process of fracture needs a crack that can be propagated across the specimen against the resistance of the material when the appropriate forcing function becomes large enough. It is an easy exercise to show (Argon, 1977) that the cracks that are necessary to bridge the gap between the technological strength levels and the cohesive strength cannot form by thermal motion under stress but must result from other processes that differ in complexity and importance between oxide glasses, metallic glasses, and glassy polymers. Particularly in glassy polymers the process of crack formation goes through a prolonged stage of stable planar cavitation called *crazing*, which can in many instances provide significant dilational strains before turning into unstable cracks. This subject deserves special attention and will be discussed in Section III.B.3.

B. CRACK FORMATION IN GLASSES

1. Crack Formation in Oxide Glasses

In oxide glasses, cracks can form (a) as a result of atmospheric corrosion producing pits, most likely at entrapped crystalline inclusions (stones) exposed at the surface (Griffith, 1920); (b) as a result of devitrification at surface inhomogeneities or at inhomogeneities in the volume where the formed crystalline inclusions produce cracks by misfit stresses (Gordon et al., 1959); and finally, (c) as a result of mechanical contact with very sharp and hard objects (Ernsberger, 1960). The areal density and severity of such cracks have been extensively studied by means of spherical indenters (Argon, 1959; Argon et al., 1960; Sucov, 1962; Matthews et al., 1976) where strength histograms are usually operationally inverted to obtain frequency distribution functions of areal densities of surface cracks with given depth (Argon, 1959a; Matthews et al., 1976). Examples of such distribution

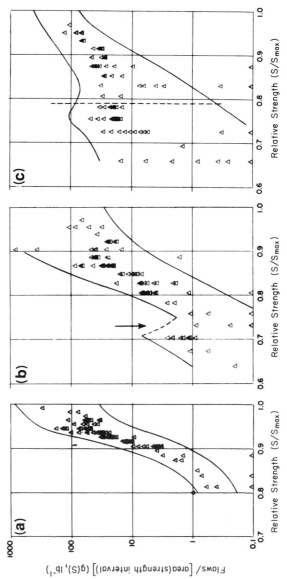

FIG. 15. Flaw density curves for: (a) chemically polished 0.01-m thick crown glass (S_{max} = 1.57 GPa); (b) chemically polished 0.01-m thick plate glass (S_{max} = 1.32 GPa); (c) mechanically polished 0.01-m thick plate glass (S_{max} = 1.02 GPa) (from Matthews *et al.* (1976), courtesy of the American Ceramic Society).

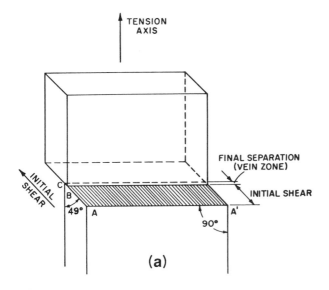

TENSION
AXIS

FINAL SEPARATION
(VEIN ZONE)

INITIAL SHEAR

INITIAL SHEAR

C

B

49°   A

90°   A'

(a)

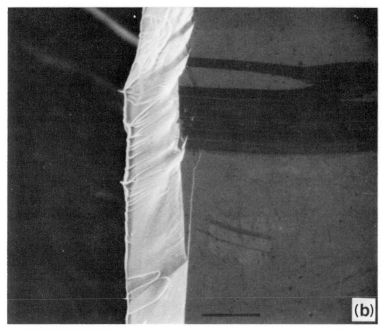

(b)

FIG. 16. Separation by shearing off in a metallic glass out of the plane of the sheet: (a) sketch; (b) sheared surface in a specimen of $Pd_{80}Si_{20}$, length of marker is 25 μm (from Megusar *et al.* (1979), courtesy of Elsevier).

functions computed from histograms (Argon *et al.*, 1960) for some typical oxide glasses with different methods of surface preparation are shown in Fig. 15 (Matthews *et al.*, 1976). Many such surface flaws have been revealed by etching with sodium vapor (Gordon *et al.*, 1959; Argon, 1959b), by lithium salt ion exchange processes (Ernsberger, 1960), or by light HF etching (Ernsberger, 1964), where either a network of surface cracks form from the preexisting sites of damage, or pits form by acid dissolution, initiating preferentially at the surface cracks and thus delineating them.

2. Crack Formation in Metallic Glasses

In metallic glasses where intense plastic shear localization can occur at low temperatures as discussed in Section II.G, very deep surface steps can form by such localization, which in many instances can lead to complete shearing off, as shown in Fig. 16 (Megusar *et al.*, 1979). In other cases, these deep surface shear steps can act as incipient cracks to lead to a unique fracture process by intrinsic cavitation in the shear band itself where the shear-induced excess free volume has reduced the plastic deformation resistance. An example is shown in Fig. 17 (Megusar *et al.*, 1979). This intrinsic cavitation process that results from a basic instability of an advancing fluid meniscus is also thought to be responsible for the production of craze matter at the tips of growing crazes in glassy polymers. This will be discussed in some detail in Section III.B.4.

Naturally, as in oxide glasses, sites of crystalline inclusions resulting from devitrification can also act as initiators of fracture in metallic glasses, but in a sense more akin to the process of cavity formation in ductile fracture. These concentrations of plastic strains around nondeformable particles set up large interfacial stresses to produce decohesion of the particle as studied extensively by Argon and co-workers (Argon *et al.*, 1975a; Argon and Im, 1975; Argon, 1976).

3. Crack Formation by Crazing in Glassy Polymers

In many homogeneous glassy polymers, the process of fracture is preceded by crazing where considerable dilational plastic strains can be accommodated in surface layers before one or more of the crazes undergo internal rupture that transforms them into cracks. These polymers tend to be brittle. Many carbon chain polymers such as polystyrene (PS), polymethylmethacrylate (PMMA), and polyphenylene oxide (PPO) fall into this category. Other polymers, such as polyethylene terephthalate (PET), and some polyimides such as resorcinol (R–R) and hydroquinone (H–H), undergo crazing in a sluggish sense, which can often be overridden by subsequent distortional plastic flow of the type discussed in Section II.E.4, and

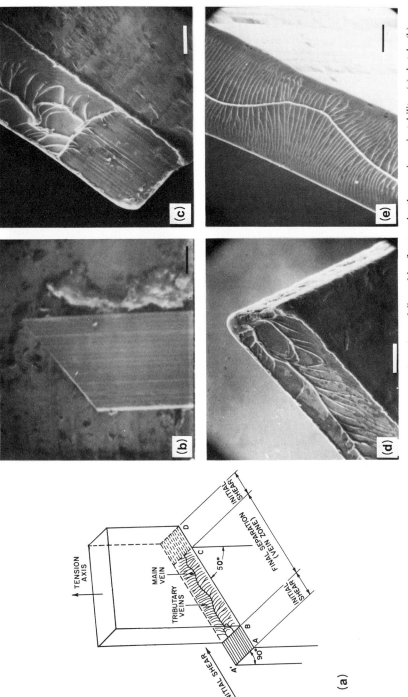

FIG. 17. Separation by initial shearing off in the plane of a metallic glass sheet, followed by fracture by the meniscus instability: (a) sketch, (b) face view of a fractured strip of $Pd_{80}Si_{20}$, length of marker is 0.1 mm; (c) and (d) opposing fracture surfaces at the reentrant corner, length of marker is 10 μm; (e) veined center portion of fracture surface, length of marker is 10 μm (from Megusar et al. (1979), courtesy of Elsevier).

113

tend not to be brittle. Still other polymers, such as polycarbonate (PC), rarely if ever craze under dry conditions without the aid of a solvent environment, and are tough. Some polymers, such as pyromellitic acid (PM; trade name, Kapton), are not known to craze and are always tough when constituted in homogeneous form (Lusignea, 1975). Although the intrinsic conditions that make some polymers crazable and others craze-resistant are not yet fully understood, considerable information is available to describe the stress and temperature dependence of craze initiation in those polymers that do craze. Sternstein and co-workers (Sternstein and Ongchin, 1969; Sternstein and Myers, 1973) have drawn attention to the fact that craze initiation from polymer surfaces is affected in a complex way by different components of the stress tensor. Both these experiments and subsequent theoretical considerations and detailed experiments by Argon (1973c, 1975) and Argon and Hannoosh (1977) have established that crazes initiate at surface irregularities in a thermally activated manner in response to a specific combined effect of the deviatoric and the negative pressure parts of the applied stress tensor. This has established that the level of the local *concentrated* deviatoric stress s [defined in Eq. (28)] governs a thermally activated coarse shear process that results directly in submicroscopic pore formation. The rate of pore production under these conditions can then be given by

$$\frac{d\beta}{dt} = \dot{\beta}_0 \exp\left\{ -\frac{1}{kT} \left[(0.15)^2 \pi \left(\frac{\mu}{s} \right) \mu \pi^3 + \alpha L^3 Y \right] \right\}, \qquad (32)$$

where β is the local volume fraction of pores, ϕ the molecular diameter, μ the shear modulus, $L = 0(4\phi)$ the radius of the critical shear patch (and the initial pore radius), Y the tensile deformation resistance, s the local deviatoric stress concentrated by the surface irregularity, and $\dot{\beta}_0$ a preexponential of order ν_G, with $\nu_G \phi L$ being the molecular segment frequency. Such submicroscopical pores are eventually expanded plastically by the average (global) negative pressure σ^0 when the mutual plastic interaction of neighboring pores becomes possible. This occurs when

$$\sigma_0 = \frac{2Y}{3} Q(s/Y,\beta) \ln(1/\beta), \qquad (33)$$

where Q, given by a cavitational "yield" locus, is the reduction of the negative pressure for plastically expanding a cavitating region when an additional deviatoric stress s is present. Combining the integral of Eq. (32) with Eq. (33) gives for the craze initiation time

$$t_{in} = (1/\dot{\beta}_0) \exp[\Delta G^*_{pore}/kT - (3\sigma_0/2Y)Q], \qquad (34)$$

where ΔG^*_{pore} is the activation-free enthalpy for pore formation given by the terms in brackets in the exponential of Eq. (32). Strictly speaking, Eq. (34) gives the time only for pore nucleation to the point where pores interact and plastic expansion into a craze nucleus begins. The overall craze initiation time in a rate-dependent material must therefore also involve the time spent in plastically expanding the pores. Unless $(\sigma_1 + \sigma_2) \to 0$, this additional time is usually quite small and can be ignored. But when $(\sigma_1 + \sigma_2) \to 0$, it becomes the dominant contribution and must be added to the term in Eq. (34). For this and other refinements, the reader is advised to consult Argon and Hannoosh (1977). Since the craze initiation time depends strongly on the severity of the surface irregularity that governs the local deviatoric stress level s, it is expected that crazes will initiate early in regions of severe stress irregularity, and as these are progressively exhausted, initiation will occur at regions of decreasing severity with increasing time. Eventually, craze initiation ceases altogether while existing crazes still continue to grow. This process of craze initiation from surfaces under multiaxial stress has been studied by Argon and Hannoosh (1977), and their measurements on PS specimens at RT and $-20°C$ at different levels of applied deviatoric stress s_0, and negative pressure σ_0 are shown in Fig. 18. Comparing these results with the theoretical model outlined above and using specific information on the distribution of surface stress concentrations partly derived from a direct measurement of surface roughness has permitted the evaluation of the parameters in Eqs. (32)–(34)*. These were found for PS to be $\dot{\beta}_0 = 4.643 \times 10^{11}$ sec^{-1}, $\alpha L^3 = 3.98 \times 10^{-28}$ m^3, and $Q = 0.0133$ for known physical constants of molecular diameter, shear modulus, yield strength, and their specific temperature dependence. Generalizing from these measurements and based on the theoretical model, Argon and Hannoosh (1977) have advanced an intrinsic biaxial craze initiation criterion for smooth surfaces, which can be given as

$$\left(\frac{\sigma_1 - \sigma_2}{2Y}\right)^2 = \left(\frac{AQ}{CQ + (\sigma_1 + \sigma_2)/2Y}\right)^{1/2} - \frac{1}{3}\left(\frac{\sigma_1 + \sigma_2}{2Y}\right)^2, \quad (35)$$

[for $(\sigma_1 + \sigma_2) > 0$], where $A = 9.31$, $C = 12.7$ at $-20°C$, and $A = 9.545$, $C = 21.23$ at $20°C$, respectively. This criterion with minor modifications [at $(\sigma_1 + \sigma_2) \to 0$] resembles that experimentally determined by Sternstein and co-workers (Sternstein and Ongchin, 1969; Sternstein and Myers, 1973). When $(\sigma_1 + \sigma_2) \to 0$, the driving force for plastic expansion of pores goes to zero and, as mentioned in connection with Eq. (34), the additional time

*The complete discussion of the distribution of the surface irregularities, how they are related to local concentrated deviatoric stresses under different states of surface biaxial stress, and how that gives rise to the specific increases in areal density of crazes under different stress levels, is outside the scope of our treatment in this chapter. For such details, the reader is referred to Argon and Hannoosh (1977).

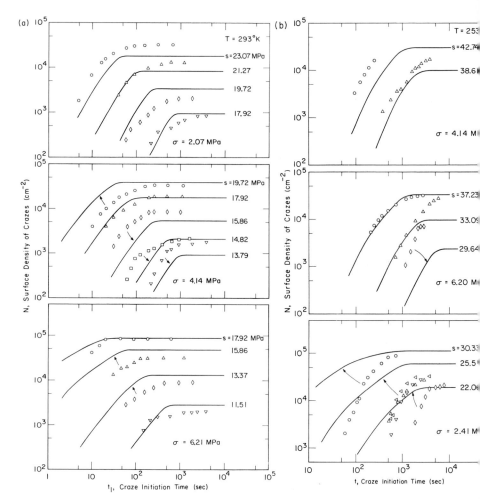

FIG. 18. Increase of craze nuclei with time in samples subjected to different combinations of global deviatoric stress and negative pressure: (a) at 293 K; (b) at 253 K. Curves computed based on theory given by Eq. (34) (from Argon and Hannoosh (1977), courtesy of Taylor and Francis).

for pore expansion needs to be considered. If such modifications are included, as discussed by Argon and Hannoosh [1977], it must be concluded that there can be no craze initiation under simple shear, and by extension in a state of biaxial surface stress where $(\sigma_1 + \sigma_2) < 0$ [see also Sternstein and co-workers (Sternstein and Ongchin, 1969; Sternstein and Myers, 1973)].

It is perhaps useful to emphasize that the model discussed here and the specific values for the parameters of the model have been established only for PS at the present, and applying them to other polymers requires caution.

Craze initiation is only the initial phase of crack formation in glassy polymers. Once initiated, crazes will grow in unoriented polymers normal to the maximum principal tensile stress at a rate governed by this stress component. As mentioned in Section III.B.2, the mechanism of craze matter production in glassy polymers and the mechanism of crack extension in metallic glasses are both based on a basic meniscus instability, which will be discussed in Section III.B.4.

When crazes become longer, certain imperfections in the craze matter can initiate local rupture in the craze matter that then propagates out in the craze plane, turning the craze quickly into a crack often resulting in final fracture as has been described in great detail by Hull and co-workers (Murray and Hull, 1970a,b; Hull, 1975). Such initiation sites of rupturing inside crazes can be readily identified by viewing the final fracture surface as is clearly illustrated in Fig. 19.

FIG. 19. Scanning electron micrograph of nuclei of rupture inside craze matter observed on fracture surface (from Murray and Hull (1970b), courtesy of John Wiley & Sons).

4. The Meniscus Instability Mechanism of Interface Convolution

Argon and Salama (1976) have drawn attention to the fact that the well-known Taylor meniscus instability that occurs when a concave interface of any fluid (whether Newtonian or non-Newtonian) is moved in the direction of its convex curvature by a suction gradient in the fluid plays a central role in both craze matter production and in the fracturing of ductile glasses. Figure 20 illustrates stages of the interface convolution process in a water meniscus made to advance between a glass plate and a cellophane tape as the latter is peeled away from the former at a critical rate (Taylor, 1950). By combining a generalized perturbation analysis of the flow of a nonlinear fluid, in a rectangular duct with the solution for the stresses and strains in the tip region of a plastically blunted crack due to McClintock (1969), Argon and Salama (1976) have determined that the steady-state wavelength λ_s and the critical opening displacement δ at the tip of the crack where the parting of the fluid due to the penetration of the finger shaped air interfaces is completed are given by

$$\lambda_s = 12 \; \pi^2 A(n)(\chi/\tau), \tag{36}$$
$$\delta = 24\pi^2 B(n)(\chi/\tau). \tag{37}$$

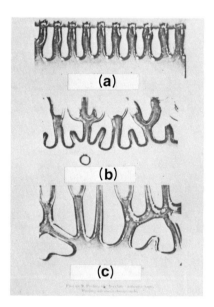

Fig. 20. Instability in an advancing fluid meniscus producing surface convolutions, (a)–(c) are stages in the development of the instability (from Taylor (1950), courtesy of Royal Society of London).

In Eqs. (36) and (37), χ is the surface free energy, τ (= $Y/\sqrt{3}$) the plastic resistance in shear, and $A(n)$ and $B(n)$ two computed functions of the strain rate sensitivity exponent n defined by the nonlinear relation between equivalent stress σ_e and equivalent strain rate $\dot{\epsilon}_e$ in a power-law representation of this connection

$$\sigma_e = {}^{3}/_{2}\eta_0(\dot{\epsilon}_e)^n, \tag{38}$$

where η_0 is a nonlinear viscosity coefficient. These functions $A(n)$ and $B(n)$ can be represented satisfactorily as

$$A(n) \simeq 1.2 \exp(4.25n), \tag{39}$$
$$B(n) \simeq 1.2 \exp(10.17n). \tag{40}$$

In Sections III.B.5 and III.C, we shall discuss the application of these results to the problem of craze matter production at the craze tip and to the ductile separation of a metallic glass by the traverse of a slightly blunted crack.

5. Craze Growth by Interface Convolution

Argon and Salama (1977) have demonstrated that rates of growth of dry crazes at the usual levels of stress cannot be explained by a mechanism of repeated pore formation at the the tips of such growing crazes. In addition, this would not result in the continuously connected air passages of craze matter. On the other hand, the interface convolution mechanism of the meniscus instability as depicted in Fig. 21 can produce readily the craze matter structure without any need for nucleating new pores in front of the craze. Furthermore, they noted that the meniscus instability provides a very satisfactory specific mechanism that also accounts for the known rates of growth of crazes.

If the craze of length $2c$ is taken as a traction transmitting elastic discontinuity as shown in Fig. 22, with a small-scale yielding zone R at the craze tip, then according to the small-scale yielding analysis the extent of this yielded zone is

$$R = c(\sigma_\infty - \sigma_0)/(Y - \sigma_\infty)^2, \tag{41}$$

where σ_∞ is the maximum principal tensile stress in the region across which the craze will grow, σ_0 the average craze face traction, and Y the tensile deformation resistance that must be constant over the region R. Argon and Salama (1977) have assumed that the average craze face traction σ_0 is governed by continued drawing of craze fibrils from the flanks and that this traction is related to the tensile deformation resistance through the effective extension ratio λ_n of these fibrils as $\sigma_0 = Y/\lambda_n$, where an inhomogeneous stress distribution factor is assumed to be absorbed into λ_n. If both σ_0 and

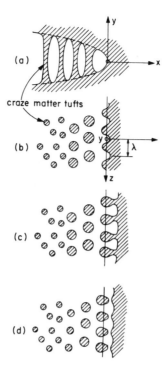

FIG. 21. Schematic drawing of craze matter production by the mechanism of interface convolution: (a) outline of craze tip; (b) cross section in craze plane across craze matter tufts; (c)–(d) advance of craze front by a completed period of interface convolution (from Argon and Salama (1977), courtesy of Taylor and Francis).

Y are strain-rate dependent according to a plastic yield condition of the type discussed in Section II.E.4, then the craze can adjust to a constant terminal velocity dc/dt with constant R for ever-increasing c, as $(\sigma_\infty - \sigma_0)$ $\rightarrow 0$ and $Y - \sigma_\infty$ approaches a constant value. Thus, in the limit, a terminal craze velocity is possible based on the steady propagation of a blunted craze tip repeatedly undergoing interface convolution. This velocity then should be (Argon and Salama, 1977)

$$\frac{dc}{dt} = \frac{\delta\dot{\epsilon}}{6}, \qquad (42)$$

where $\dot{\epsilon}(= \dot{\gamma}/\sqrt{3})$ is the tensile plastic strain rate determinable from Eq. (21) by setting $\sigma = Y/\sqrt{3}$ and $\dot{\gamma} = \dot{\epsilon}\sqrt{3}$, and where δ is the steady-state craze opening displacement given by Eq. (37), while the factor 6 in the denominator comes from the details of the theoretical model. Thus, intro-

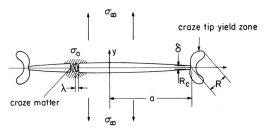

FIG. 22. Craze as a traction transmitting discontinuity; showing craze tip plastic zone, craze tip opening displacement, and craze matter producing craze face traction σ_0 (from Argon and Salama (1977), courtesy of Taylor and Francis).

ducing these specific developments, the terminal craze velocity is obtained as

$$\frac{dc}{dt} = \frac{D}{(\lambda_n \sigma_\infty / \hat{Y})} \exp\left\{ -\frac{A}{T}\left| 1 - \left(\frac{\lambda_n \sigma_\infty}{Y}\right)^{5/6}\right| \right\}, \tag{43}$$

where

$$D = 24\pi\sqrt{3}\, B(n)\chi\dot{\gamma_0}/\hat{Y}, \tag{44}$$
$$\hat{Y} = [0.133/(1 - \nu)]\mu, \tag{45}$$

$$A = \frac{3\pi\mu\omega^2 a^3}{16(1 - \nu)k} \tag{46}$$

and where $\dot{\gamma_0}$ is the preexponential factor in Eqs. (8) and (21), with all other terms remaining as defined in Section II.E.4. In addition, a measure of the mean spacing between craze tufts is obtainable from the steady-state wavelength λ of the convoluting interface as

$$\lambda = 12\pi^2\, A(n)\sqrt{3}\, \chi/Y. \tag{47}$$

Experimental measurements of craze growth broadly confirm the predictions of the theoretical model. The measured terminal craze velocities in two types of PS and a commercial PMMA at both 20°C and −20°C are plotted in Fig. 23 as a function of the reduced tensile stress $\sigma_\infty/\hat{Y}$. As Argon and Salama discuss, the straight lines passing through the data points are in agreement with the theoretical model after some minor adjustments are made for the strength-differential effect, and imply that the effective extension ratio of the craze matter tufts is between 1.5 and 2.5. Since the actual extension ratios measured from electron micrographs are in the range of 4.5–5, a shoulder stress concentration factor of 2–3 is implied, which is also consistent with expectations of the levels of pressure generated in the

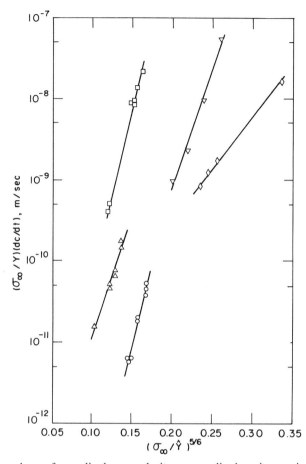

FIG. 23. Dependence of normalized craze velocity on normalized maximum principal tensile stress at two temperatures for two grades of PS and a PMMA Δ, PS/293 K; ∇, PS/253 K; □, Dow 686 PS/293 K; ◇ Dow 686 PS/253 K; ○, PMMA/293 K. (After Argon and Salama (1977).)

shoulder regions of a drawing acute neck (Argon *et al.*, 1975b). For the actual quantities that were used for the evaluation of Eqs. (43)–(46) in matching the theoretical model to the experimental measurements, the reader is referred to Argon and Salama (1977).

In addition, the theory predicts and the multiaxial crazing experiments confirm that crazes in unoriented polymers grow only in response to the maximum principal tensile stress in the field. This emphasizes that in a state of simple shear where crazes may not be able to initiate, they can nevertheless readily grow. This is a matter of considerable importance since the usual craze initiation locus of the type given by Eq. (35) or by the similar forms published by Sternstein and co-workers are often wrongly assumed to hold for the entire crazing process.

It is essential to emphasize that the crazing process discussed here is that which occurs in the absence of any solvents. When solvents are present, crazing is more or less accelerated. This problem of crazing is different enough to require special attention. It is another subject that falls outside the scope of this chapter. The interested reader is referred to Kramer and Bubeck (1978) for a discussion of solvent crazing.

As already mentioned, when crazes become long, either at the initial surface irregularity or at one or more entrapped inhomogeneities in the craze matter, actual rupturing of a bundle of neighboring craze tufts can occur. This acts as a true crack nucleus that propagates inside the craze out to its extremities, and converts that craze to a crack pushing a narrow region of craze at its tip acting as a planar plastic zone.

C. CRACK PROPAGATION IN GLASSES

1. Oxide Glasses

Griffith (1920) in his classical work on fracture was first to demonstrate that cracks propagate in oxide glasses when the rate of release of elastic energy equals the rate of production of the energy of fresh surfaces or, as Orowan (1934) pointed out somewhat later, when the concentrated stress at the tip of the often atomically sharp crack reaches the ideal cohesive strength σ_i of the glass. Since oxide glasses are potentially brittle solids according to the basic classification of Kelly et al. (1967), this propagation is not accompanied by any significant amount of plastic deformation. Hence, the critical stress intensity factor K_{Ic} becomes

$$K_{Ic} = (2E\chi)^{1/2} = \sigma_i(a\pi/2)^{1/2}, \tag{48}$$

where E is Young's modulus, χ the surface energy, and a the atomic radius. In most inorganic glasses, the surface energy is of order 0.5 J/m² (Griffith, 1920) and Young's modulus of order 70 GPa. This makes the critical stress intensity factor K_{Ic} of order 0.3 MPa m$^{1/2}$, which is very close to the value 0.28 MPa m$^{1/2}$ measured by Griffith (1920) on precracked tubes and spherical bulbs of a conventional soda glass of 0.692 SiO_2, 0.12 K_2O, 0.009 Na_2O, 0.118 Al_2O_3, 0.045 CaO, and 0.009 MnO. Most experiments since the time of Griffith have confirmed this picture.

2. Metallic Glasses

As we have already shown in Fig. 17, fracture occurs in metallic glasses by an intrinsic cavitation process involving the meniscus instability. Deep surface offsets at shear bands act as the initiating sites from which cracks propagate inward, usually, but not necessarily always, along the shear bands where the deformation-induced excess free volume has lowered the plastic resistance. In this fracture process, the basic mechanism of separation is ductile rupture along the steady-state ridges between the finger-shaped protrusions at the convoluted crack tip penetrating almost mono-lithically into the region ahead of the crack tip—in a manner shown in Fig. 20. Here again the development of Argon and Salama (1976) for the parameters of the convoluted meniscus interface of a nonlinear fluid permits writing down the steady-state crack opening displacement δ [Eq. (37)], the fracture surface ridge spacing λ, [Eq. (36)] and the fracture toughness K_{Ic} as

$$K_{\text{Ic}} \equiv \left(\frac{\delta \pi YE}{\alpha} \right)^{1/2} = \left(\frac{24\pi^3 \sqrt{3} \, B(n)\chi E}{\alpha} \right)^{1/2}, \tag{49}$$

where $\alpha \approx 2.7$ is a numerical constant giving the ratio of the critical crack opening displacement to the product of the tensile yield strain and the critical plastic zone size. In metallic glasses at low temperatures where the strain rate sensitivity of the flow stress is nil, $B(n) \approx 1.2$. In addition most metallic glasses have Young's moduli of order 140 GPa, tensile yield stresses of order 2.5 GPa, and surface energies of order 2 J/m^2, which gives for the fracture toughness K_{Ic} about 10 MPa m$^{1/2}$ and for the ridge spacing on the fracture surface $\approx 2.5 \times 10^{-7}$ m. Davis (1976) has performed a number of plane strain fracture toughness experiments on samples of metallic glass precracked by fatigue crack propagation. He has found these fracture toughnesses to range from a low of 9.5 MPa m$^{1/2}$ for a glass of $Ni_{49}Fe_{29}P_{14}B_6Si_2$ to a high of 12.65 MPa m$^{1/2}$ for the strongest glass of $Fe_{80}B_{20}$. These values are in remarkably good agreement with the prediction of the meniscus convolution model of crack propagation.

3. Glassy Polymers

In glassy polymers, when crazes are transformed into supercritical cracks, or when other inclusions or large-scale surface irregularities act as supercritical cracks, catastrophic fracture may follow. In their growth, such cracks will be blunted by inelastic deformation at the crack tip that can be a mixture of plastic flow and additional crazing occurring in a zone having the dimensions R given by the small-scale yielding theory as

$$R = c(\sigma_\infty / Y)^2, \tag{50}$$

where the symbols have their previously defined meaning. Crack propagation occurs when the crack opening displacement δ reaches a critical value, which in this case is most often governed by continued rupture in the parent craze that gave rise to the initial crack, or a series of ruptures in adjoining crazes bridged by some plastic flow and tearing. Thus, the fracture toughness is again governed by the first equality of Eq. (49) except that in this case the interpretation of the crack opening displacement is not so clear.

Williams (1977) has examined the plane strain fracture condition of a number of glassy polymers and has found them to be governed by a critical stress intensity factor criterion given by Eq. (29) as shown in Figs. 24–26 for PMMA, PS, and PC, respectively. Evaluation of the slopes of these curves has given the critical stress intensity factors shown in Fig. 27. It is clear that, in the more brittle polymers of PMMA and PS, the fracture toughness is relatively temperature independent, while in PC with a much higher fracture toughness this value is also quite temperature dependent. Evaluation of the critical crack opening displacements by Eq. (49) gives values of 10^{-5} m for PS, 4×10^{-5} m for PMMA, and 2.5×10^{-4} m for PC, respectively. These are all much larger than the typical craze thicknesses of 10^{-7}–10^{-6} m and indicate that crack propagation in glassy polymers is normally not governed by the splitting of a single craze but that rupture in many parallel crazes is involved with bridging tears between these. In PC, which does not craze readily, eventual tearing is preceded by very considerable distortional crack tip blunting.

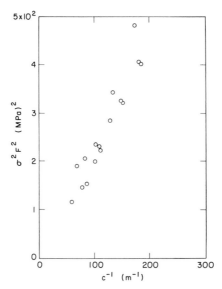

FIG. 24. Dependence of fracture stress on initial crack length in PMMA $K_{Ic} = 2.64$ MPa $\text{m}^{1/2}$ (from Williams (1977), courtesy of *Polymer Eng. Sci.*).

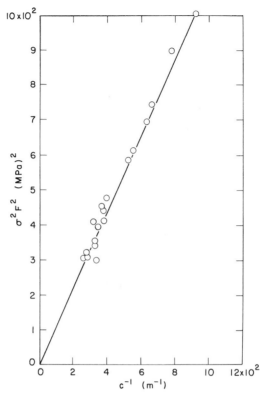

Fig. 25. Dependence of fracture stress on initial crack length in PS $K_{Ic} = 1.86$ MPa · m$^{1/2}$ (from Williams (1977), courtesy of *Polymer Eng. Sci.*).

IV. Other Aspects of Fracture in Glasses

In the preceding sections of this chapter, only the basic mechanisms of inelastic deformation and fracture of glasses could be covered. There are many other important aspects of this subject that had to be excluded. In view of this, it is useful to guide the interested reader to some key references where extensive coverage of some of these excluded topics can be found.

Nearly all polymers exhibit large amounts of recoverable nonlinear viscoelastic strains accompanying the semipermanent plastic deformations that were discussed in Section II.D.4. Although the basic mechanism of nonlinear viscoelasticity must be the same as that of plastic flow, its kinetics are complicated by the molecular alignment that accompanies it. The usual treatments of the subject have emphasized phenomenological and operational aspects rather than mechanistic ones. An adquate review of

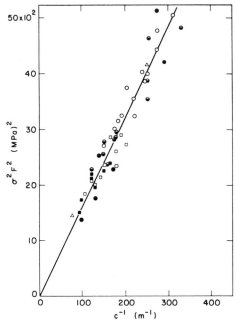

FIG. 26. Dependence of fracture stress on initial crack length and loading rate in PC K_{Ic} = 6.9 MPa · $m^{1/2}$. Machine notched: ○, 0.5 cm/min; ◒, 0.05 cm/min. Razor notched: ●, 0.5 cm/min; △, 5.0 cm/min; □, 20.0 cm/min; ■, 50.0 cm/min (from Williams (1977), courtesy of *Polymer Eng. Sci.*).

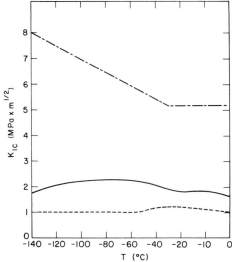

FIG. 27. Dependence of critical stress intensity factor K_{Ic} for fracture on temperature in PMMA (——), PS (---), and PC (—·—) (from Williams (1977), courtesy of *Polymer Eng. Sci.*).

these developments can be found in the general treatise of Gittus (1975). Some viscoelastic behavior, both linear and nonlinear, is also exhibited by oxide glasses and metallic glasses. Reference was already made to the study of Li and Uhlmann (1970a,b) of the high-stress behavior of some oxide glasses.

Although oxide glasses cannot undergo much plastic strain and plastic flow in metallic glasses produces no hardening, when glassy polymers are deformed they undergo orientation hardening. The behavior of oriented polymers is an extensively explored subject of considerable technological importance. A recent review of many aspects of oriented polymers including their deformation and fracture has been assembled by Ward (1975).

The fatigue fracture of glasses under cyclic deformation has been investigated in considerable detail. In oxide glasses, cyclic fatigue is almost exclusively a result of progressive stress corrosion cracking by water vapor occurring during the tensile portions of each cycle. This is discussed by Freiman in Chapter 2. In metallic glasses, crack growth occurs by cyclic plastic crack tip distortions. Davis (1978) has reviewed the available information on fatigue in metallic glasses and has reported that such growth rate obeys a second-power dependence on the amplitude of the stress intensity factor, but finds some divergence in the results of some other authors. Fatigue in glassy polymers has also been extensively investigated. Adequate accounts of the cyclic response of polymers have been given by Beardmore and Rabinowitz (1975) and the phenomena of crack propagation have been summarized by Manson and Hertzberg (1973) and Kim et al. (1977).

Finally, the subject of the mechanical behavior of glasses in heterogeneous mixtures and composites is very large with references being vast. As a beginning, the reader is referred to the many treatises on composite materials, particularly those edited by Broutman and Krock (1974). An introduction to the subject of rubber-reinforced polymers that is of ever-increasing importance can be had in the recent treatise of Bucknall (1977).

Appendix: Plastic Flow Mechanism of Oxide Glasses

For simplicity, we consider a simple linear interlayer shear resistance given by

$$\frac{\tau}{\hat{\tau}} + \frac{2\gamma}{\gamma_0} = 1 \qquad (A.1)$$

between $0 \leq \gamma \leq \gamma_0$, with vertical rises at $\gamma = 0$ and $\gamma = \gamma_0$. Then for any local homogeneous shear γ in the region of volume Ω_f, the Helmholtz free energy is

$$\Delta F = \hat{\tau}\Omega_f\gamma_0\{(\gamma/\gamma_0) [1 - \gamma/\gamma_0]\}. \tag{A.2}$$

The elastic strain energy $\Delta\mathscr{E}$ due to a homogeneous shear γ_0 over only the volume Ω_f, with no imposed shear outside the region is

$$\Delta\mathscr{E} = B(\hat{\tau}\Omega_f\gamma_0)_-(\gamma/\gamma_0)^2, \tag{A.3a}$$

$$B = [(7 - 5\nu)/30(1 - \nu)] \mu\gamma_0/\hat{\tau}. \tag{A.3b}$$

The work done by the external stress σ resulting from this shear is

$$\Delta W = \hat{\tau}\Omega_f\gamma_0(\sigma/\hat{\tau})(\gamma/\gamma_0). \tag{A.4}$$

The activation configuration is obtained for constant stress by maximizing the free enthalpy ΔG as

$$\left(\frac{\partial\Delta G}{\partial(\gamma/\gamma_0)}\right)_{\sigma/\hat{\tau}} = \left[\frac{\partial}{\partial(\gamma/\gamma_0)}(\Delta F + \Delta\mathscr{E} - \Delta W)\right]_{\sigma/\hat{\tau}} = 0, \tag{A.5}$$

which gives

$$(\gamma/\gamma_0)^* = (1 - \sigma/\hat{\tau})/2(1 - B), \tag{A.6}$$

and substitution of $(\gamma/\gamma_0)^*$ into ΔG gives the activation-free enthalpy

$$\Delta G^* = [\hat{\tau}\Omega_f\gamma_0/4(1 - B)] [1 - \sigma/\hat{\tau}]^2. \tag{A.7}$$

AKNOWLEDGMENT

The author's research in polymers has been supported primarily by the National Science Foundation under Grants GH-40467, DMR-73-02440-A01, and DMR-77-22753, while the research on metallic glasses has been supported by the MRL Division of the National Science Foundation through the Center for Materials Science and Engineering at MIT under Grants DMR-76-80895 and DMR-76-80895-A02.

References

Argon, A. S. (1959a). *Proc. R. Soc. London Ser. A* **250**, 482.
Argon, A. S. (1959b). *Proc. R. Soc. London Ser. A* **250**, 472.
Argon, A. S. (1969). Unpublished observations.
Argon, A. S. (1973a). *Philos. Mag.* **28**, 839.
Argon, A. S. (1973b). *In* "The Inhomogeneity of Plastic Deformation," p. 161. American Society of Metals, Metals Park, Ohio.
Argon, A. S. (1973c). *J. Macromol. Sci. Phys.* **B8**, 573.
Argon, A. S. (1975). *Pure Appl. Chem.* **43**, 247.
Argon, A. S. (1976). *J. Eng. Mater. Technol.* **98**, 60.
Argon, A. S. (1977). *In* "Surface Effects in Crystal Plasticity" (R. M. Latanision and J. F. Fourie, eds.), p. 383. Noordhoff, Leyden.
Argon, A. S. (1979). *Acta Metall.* **27**, 47.
Argon, A. S., and Bessonov, M. I. (1977a). *Philos. Mag.* **35**, 917.
Argon, A. S., and Bessonov, M. I. (1977b). *Polym. Sci. Eng.* **17**, 174.
Argon, A. S., and Hannoosh, J. G. (1977). *Philos. Mag.* **36**, 1195.

Argon, A. S., and Im, J. (1975). *Metall. Trans.* **6A**, 839.
Argon, A. S., and Kuo, H. Y. (1979). *Mater. Sci. Eng.* **39**, 110.
Argon, A. S., and Salama, M. M. (1976). *Mater. Sci. Eng.* **23**, 219.
Argon, A. S., and Salama, M. M., (1977). *Philos. Mag.* **36**, 1217.
Argon, A. S., Hori, Y., and Orowan, E. (1960). *J. Am. Ceram. Soc.* **43**, 86.
Argon, A. S., Andrews, R. D., Godrick, J. A., and Whitney, W. (1968). *J. Appl. Phys.* **39**, 1899.
Argon, A. S., Im, J., and Safoglu, R. (1975a). *Metall. Trans.* **6A**, 825.
Argon, A. S., Im, J., and Needleman, A. (1975b). *Metall. Trans.* **6A**, 815.
Bauwens, J. C., (1970). *J. Polym. Sci.*, A-2, **8**, 893.
Bauwens-Crowet, C., Bauwens, J. C., and Homes, G. (1969). *J. Polym. Sci. Polym. Phys. Ed.* **7**, 735.
Bauwens-Crowet, C., Bauwens, J. C., and Homes, G. (1972). *J. Mater. Sci.* **7**, 176.
Beardmore, P., and Rabinowitz, S. (1975). *In* "Treatise on Materials Science and Technology" (C. McMahon, ed.), Vol. 6, p. 267. Academic Press, New York.
Bernal, J. D. (1964) *Proc. R. Soc. London Ser. A* **280**, 299.
Bowden, P. B., and Henshall, J. L., unpublished, quoted in Argon (1973).
Bowden, P. B., and Raha, S. (1970). *Philos. Mag.* **22**, 463.
Bowden, P. B., and Raha, S. (1974). *Philos. Mag.* **29**, 149.
Bragg, L., and Lomer, W. M. (1949). *Proc. R. Soc. London Ser. A* **196**, 171.
Bragg, L., and Nye, J. F. (1947). *Proc. R. Soc. London Ser. A* **190**, 474.
Bridgman, P. W., and Simon, I. (1953). *J. Appl. Phys.* **24**, 405.
Broutman, L. J., and Krock, R. H. (eds.) (1974). "Composite Materials" (multi-volume treatise). Academic Press, New York.
Bucknall, C. B. (1977). "Toughened Plastics." Applied Science Publ., Essex, England.
Davis, L. A. (1976). *In* "Rapidly Quenched Metals" (N. J. Grant and B. C. Giessen, eds.), Sect. 1, p. 369. MIT Press, Cambridge, Massachusetts, 1976.
Davis, L. A. (1978). *In* "Metallic Glasses" p. 190. American Society of Metals, Metals Park, Ohio.
Douglas, R. W. (1958). *J. Soc. Glass Technol.* **42**, 145.
Ernsberger, F. M. (1960). *Proc. R. Soc. London Ser. A* **257**, 213.
Ernsberger, F. M. (1962). *Adv. Glass Technol.* (*Proc. Int. Conf. Glass, 6th, Washington, July*), Vol. 1, p. 511. Plenum Press, New York.
Ernsberger, F. M. (1968). *J. Am. Ceram. Soc.* **51**, 545.
Eshelby, J. D. (1957). *Proc. R. Soc. London Ser. A* **241**, 376.
Gittus, J. (1975). "Creep, Viscoelasticity, and Creep Fracture in Solids," p. 403. Applied Science Publ., Essex, England.
Gordon, J. E., Marsh, D. M., and Parratt, M. E. M. L. (1959). *Proc. R. Soc. London Ser. A* **249**, 65.
Griffith, A. A. (1920). *Philos. Trans. R. Soc.* **A221**, 163.
Griffith, A. A. (1924). *Proc. Int. Congr. Appl. Mech. 1st, Delft* p. 55.
Haward, R. N. (1973). *In* "The Physics of Glassy Polymers," (R. N. Haward ed.) pp. 24, 42. Wiley, New York.
Hay, A. S. (1969). *Encycl. Poly. Sci. Technol.* **10**, 101.
Hillig, W. B. (1962). *In* "Modern Aspects of the Vitreous State" (J. D. Mackenzie, ed.), Vol. 2, p. 152. Butterworths, London,
Hull, D. (1975). *In* "Polymeric Materials," p. 487. (American Society of Metals, Metals Park, Ohio.
Illers, K. H., and Breuer, H. (1963). *J. Colloid Sci.* **18**, 1.
Irwin, G. R. (1948). *In* "Fracturing of Metals," p. 147. American Society of Metals, Metals Park, Ohio.

Kelly, A., Tyson, W. R., and Cottrell, A. H. (1967). *Philos. Mag.,* **15**, 567.

Kerkhof, F., and Richter, H. (1969). *Glastech. Ber.* **42**, 129.

Kim, S. L., Skibo, M., Manson, J. A., and Hertzberg, R. W. (1977). *Polym. Eng. Sci.* **17**, 194.

Kocks, U. F., Argon, A. S., and Ashby, M. F. (1975). Thermodynamics and kinetics of slip, *Prog. Mater. Sci.* **19**, (1975).

Kramer, E. J., and Bubeck, R. A. (1978). *J. Polym. Sci. -Phys.* **16**, 1195.

Li, J. C. M., and Gilman, J. J. (1970). *J. Appl. Phys.* **41**, 4248.

Li, J. H., and Uhlmann, D. R. (1970a). *J. Non-Crystall. Solids* **3**, 127.

Li, J. H., and Uhlmann, D. R. (1970b). *J. Non-Crystall. Solids* **3**, 205.

Lusignea, R. W. (1975). Crazing and Fracture in Polyimide and Polyethylene-Terephthalate. S. M. Thesis, MIT, Cambridge, Massachusetts.

Manson, J. A., and Hertzberg, R. W. (1973). *Crit. Rev. Macromol. Sci.* **1**, 433.

Marsh, D. M. (1964a). *Proc. R. Soc. London Ser. A* **279**, 420.

Marsh, D. M., (1964b). *Proc. R. Soc. London Ser. A* **282**, 33.

Matthews, J. R., McClintock, F. A., and Shack, W. J. (1976). *J. Am. Ceram. Soc.* **59**, 304.

McClintock, F. A. (1969). *In* "Physics of Strength and Plasticity" (A. S. Argon, ed.), p. 307. MIT Press, Cambridge, Massachusetts. (1969).

McClintock, F. A., and Argon, A. S. (1966). "Mechanical Behavior of Materials," p. 495. Addison-Wesley, Reading, Massachusetts.

Megusar, J., Argon, A. S., and Grant, N. J. (1979). *Mater. Sci. Eng.* **38**, 63.

Murray, J., and Hull, D. (1970a). *J. Polym. Sci. Polym. Lett. Ed.* 159.

Murray, J., and Hull, D. (1970b). *J. Polym. Sci. Polym Phys. Ed.* **8**, 583.

Neuhäuser, H. (1978). *Scr. Metall.* **12**, 471.

Orowan, E. (1934). *Z. Kristallogr.* **89**, 327.

Orowan, E. (1949). *Rep. Prog. Phys.* **12**, 185.

Orowan, E., *Proc. Nat. Congr. Appl. Mech.* p. 453. American Society Mechanical Engineers, 1952.

Pampillo, C. A. (1975). *J. Mater. Sci.* **10**, 1194.

Pampillo, C. A., and Chen, H. S. (1974). *Mater. Sci. Eng.* **13**, 181.

Paris, P. C., and Sih, G. C. (1965). *In* "Fracture Toughness Testing and its Applications," STP-381, p. 30. American Society Testing and Materials, Philadelphia, Pennsylvania.

Peter, K. (1964). *Glastech. Ber.* **37**, 333.

Peyches, I. (1952). *J. Soc. Glass Technol.* **36**, 164.

Raghava, R., Caddell, R. M., and Yeh, G. S. Y. (1973). *J. Mater. Sci.* **8**, 225.

Robertson, R. E. (1966). *J. Chem. Phys.* **44**, 3950.

Robertson, R. E. (1968). *Appl. Polym. Symp.* **7**, 201.

Rooke, D. P., and Cartwright, D. J. (1976). "Compendium of Stress Intensity Factors." HM Stationery Office, London.

Rudd, J. F., and Gurnee, E. F. (1957). *J. Appl. Phys.* **28**, 1096.

Spaepen, F. (1977). *Acta Metall.* **25**, 407.

Sucov, E. W. (1962). *J. Am. Ceram. Soc.* **45**, 214.

Sternstein, S. S., and Myers, F. A. (1973). *J. Macromol. Sci. Phys.* **B8**, 539.

Sternstein, S. S., and Ongchin, L. (1969). *Polym. Preprint* **10**, 117.

Tada, H., Paris, P. C., and Irwin, G. R. (1973). "The Stress Analysis of Cracks Handbook. Del Research Corp., Hellertown, Pennsylvania.

Taylor, G. I. (1950). *Proc. R. Soc. London Ser. A* **201**, 192.

von Mises, R. (1913). *Goettinger Nachr. Math-Phys. Klasse* 582.

Ward, I. M. (ed.) (1975). "Structure and Properties of Oriented Polymers." Halsted Press, New York.

Ward, I. M., and Foot, J., unpublished, quoted in Argon (1973).

Warren, B. E. (1937). *J. Appl. Phys.* **8**, 645.

Warren, B. E., Krutter, H., and Morningstar, O. (1936). *J. Am. Ceram. Soc.* **19**, 202.
Whitney, W. (1963). *J. Appl. Phys.* **34**, 3633.
Whitney, W., and Andrews, R. D., (1967). *J. Polym. Sci. C*, **16**, 1981.
Williams, J. G. (1977). *Polym. Eng. Sci.* **17**, 144.
Yannas, I. V., and Lunn, A. C. (1975). *Polym. Preprint.* **16**, 564.

CHAPTER 4

Techniques of Strengthening Glasses

F. M. Ernsberger

PPG INDUSTRIES, INC.
PITTSBURGH, PENNSYLVANIA

I. Introduction

The purpose of this chapter is to provide an overview of the concept of glass strengthening. Other chapters will supply details on those technologies that have matured to commercial significance. Included in this overview will be some of the more speculative notions about possible approaches to the strengthening of glass, as well as reference to some technologies that have not yet seen significant practical application.

The reader may wish to consult other recent reviews in this field. LaCourse (1972) includes a substantial section on the fundamentals of strengthening in a general review on strength of glass. Shoemaker (1969) treats practical aspects of the subject, and there is an earlier review by the present author (Ernsberger, 1966).

133

II. Surface Damage

There is an interesting paradox in the fact that the strongest brittle materials are stronger than the strongest ductile ones. The contrast is most striking when one compares the same material in brittle and ductile forms. Single-crystal whiskers of iron, for example, are enormously stronger than polycrystalline samples of the same metal. The contrast reminds us that the ability to yield in shear, which is the property that imparts ductility, is after all a form of failure. Thus the structural toughness and reliability that we associate with the ductile metals is purchased at the expense of ultimate tensile strength.

Ductility in a crystalline material (excluding molecular crystals from the discussion) is a consequence of the presence of dislocations that exist in a thermally activated state at the temperature of use. The application of a stress organizes the motion and/or the multiplication of these dislocations in such a way that the stress is relieved. Only in the case of extremely small monocrystalline specimens that have no dislocations, such as the metallic whiskers mentioned above, do we encounter the inherent strength of a metal such as iron; and without dislocations, the material is necessarily brittle.

In the case of oxide glasses, the amorphous structure precludes the existence of dislocations in the ordinary sense. Even if dislocations did exist, it can be inferred from the brittleness of related crystalline species such as quartz that they would be immobile at ordinary ambient temperatures. According to our present understanding of the nature of the partly covalent, partly ionic bonding of the oxide glass structure, there is no way to accomplish what nature has already provided in many of the metals; that is, to barter some of the superabundant intrinsic strength of glasses for a measure of ductility. This being the case, the only alternative is to study carefully the nature of brittle fracture in order to acquire a degree of control over its initiation and propagation.

A fundamental contribution was made by Griffith (1921) many years ago. He postulated the existence of minute, cracklike flaws in all materials. The stress-concentrating effect of these flaws is unimportant in ductile materials because a minor amount of flow in the region of the crack tip relieves the local over-stressed condition. However, brittle materials do not have this option, so crack propagation and failure by fracture will occur at a low value of mean stress.

Griffith originally assumed these cracks to be distributed throughout the volume of materials, as no doubt they are for polycrystalline metals and ceramics. In the case of glass, however, subsequent research has shown that the microcracks exist only at the surface. Indeed it seems almost intuitively obvious that microcracks cannot exist in the interior, in view of the known continuity between the liquid and glassy states.

Many suggestions have been made concerning the manner in which surface microcracks originate. Spontaneous crack formation from cooling stresses, chemical effects, and many other concepts have been cited. However, convincing evidence now exists that indicates that the vast majority of strength-impairing surface structures are microfractures arising from accidental mechanical damage that occurs during manufacture, processing, and use (Ernsberger, 1962).

Another thing we know about brittle fracture is the fact that it is a low-energy process. The cleavage energy of a soda-lime glass in dry nitrogen at room temperature is about 3800 ergs/cm^2 (Wiederhorn, 1969), whereas that of mild steel at a temperature just below the ductile–brittle transition is of the order of 10^6 ergs/cm^2. This vast discrepancy is credibly assigned to the existence of a sizable zone of plastic deformation at the tip of a crack propagating in iron, whereas in glass there is none. Perhaps "none" is too strong a word, because the cleavage energy of glass substantially exceeds the 500–600 ergs/cm^2 that is regarded as a reasonable value for the thermodynamically defined surface free energy of glass at room temperature. A plastic zone of about 50 Å diameter could account for this, but it appears more likely that the cleavage of glass inevitably produces a high-energy surface, bearing a high concentration of free radicals and unscreened ions.

Still another characteristic feature of fracture initiation in glass is the fact that it is sensitive to chemical influences from the ambient atmosphere (Charles, 1958). In other words, glass is subject to stress–corrosion fatigue. The usual corrosive reagent is water, as liquid, vapor, or aqueous solutions. It follows that fracture initiation at constant stress is time dependent. In a vacuum or dry-gas ambient, this time dependency is much reduced, and at very low temperatures it altogether disappears.

Temperature affects the strength of glass in subtle and complex ways. Its effect on the stress–corrosion reaction tends to obscure its more fundamental role in the activation of bond rupture. In vacuum or a dry atmosphere where stress–corrosion is presumably absent, the expected downward trend of strength with increasing temperature is observed, but experimentally the relation is sigmoidial rather than monotonic. This suggests that some sort of diffusive internal rearrangement is involved in the initiation of fracture, particularly in the temperature range 150–450 K (Ernsberger, 1969).

III. Avoidance of Surface Damage

It must be evident from the discussion in the preceding section that the technology of strengthening glass converges on a single issue: what to do about surface damage. One approach that might be contemplated is to avoid surface damage altogether. In principle this is the best possible approach, because the strength of damage-free (pristine) glass cannot be improved upon. The main difficulty with this approach is implied by the axiom that a chain is only as strong as its weakest link. It is not enough to have a *nearly* perfect surface; it must be *absolutely* perfect if the inherent strength of glass is to be realized. Moreover, this state of perfection must be maintained for as long as high-strength performance is required.

The stringency of these requirements is obviously much reduced if only small areas of glass must be perfect, and if the nature of the use is such that permanent protection is ensured. These two conditions are met in the case of fiberglass-reinforced plastics. The small-area, permanent-protection conditions also contribute significantly to the useful strength of glass containers. The stresses produced by internal pressure or by external impact are highest on the interior surface, and in a narrow-necked container this surface tends to preserve its initial pristine character.

In the case of flat glass, however, it is quite difficult to take advantage of pristine strength. It has been proposed to laminate sheets of pristine glass between layers of plastic or of ordinary glass (Van Laethem, 1972). However, the performance advantage of such constructions hardly justifies the extra cost unless there are special circumstances. In particular, the use of plastic outer plies to protect a glass surface sacrifices one of the most useful properties of glass, namely, its ability to maintain acceptable optical clarity despite the scratches and abrasions that result from normal use and periodic cleaning.

There is also the question of edge strength. Conventional methods of cutting flat glass, whether sliding diamond or rolling wheel, leave a dense trail of microscopic fissures along the cut edge. These fissures frequently have a worse strength-impairing effect than that associated with the normal level of surface damage. There is little point in improving the perfection of the major surfaces of a sheet of glass if the strength is controlled by the condition of the edges.

It is not particularly difficult to produce flat glass with a pristine surface. Drawing processes, with the exception of the Colburn process, can be modified in such a way that at least a portion of the ribbon is untouched (Ernsberger, 1967). The float process probably could also be so adapted, although it would be more difficult. It has also been proposed (Charles, 1966) that a pristine surface could be generated and simultaneously buried out of harm's way by treating flat glass with saturated steam under conditions such that a layer of hydrothermally generated microcrystalline material covers the surface. Optical transparency would of course be sacrificed.

A related concept (Stookey, 1970) proposes the use of certain special compositions that, when treated with steam, are converted to any desired depth into a rubbery material that protects the brittle interior from damage.

The rubbery surface, because of its low modulus, is subject to very little stress and hence is relatively immune to strength impairment by surface damage. There is no indication that either of these steam-treatment processes has had any commercial use.

The surface-protection concept is technologically very important in the container industry. A freshly blown bottle is extremely susceptible to abrasion, owing to the extraordinarily clean condition of the surface. The coefficient of friction between clean glass surfaces is extremely high; welding apparently occurs at points of contact, and subsequent rupture of these welds severely damages the glass. This damage can be largely avoided by treatment of the outside surface with lubricants. A brief review of the theory and practice of this technology is given by Scholes (1971). Lubricant coatings may be applied at either the hot or the cold end of the annealing lehr, or both. At the hot end it is common practice to spray the bottles with solutions of tin or titanium compounds or silicones; at the cold end, oily substances such as oleic acid are used.

IV. Removal of Surface Damage

Another thing one can do about surface damage is to remove it. Thorough removal of surface damage should bring the strength back to that of the pristine state. The problem of maintaining the high-strength state is of course the same as discussed in the preceding section.

A. MICROCRACK HEALING

Most surface damage, as noted in Section II, consists of microcracks. The mating surfaces of these cracks are so close together that they are optically in contact, and hence invisible in the microscope. This being the case, it would seem that means could be found to bring about a "healing" or reunion of the microcrack surfaces, thereby restoring the strength of the glass.

There is one reported experimental observation (Wiederhorn, 1970) of a partial spontaneous healing of freshly formed cracks in glass, but as a practical matter, conditions for the healing of microcracks have not been found. Perhaps this is not surprising because a fresh glass fracture surface irreversibly chemisorbs O_2, CO_2, and H_2O (Antonini, 1969), making the ruptured bonds unavailable for recombination. Moreover, the nonrepetitive structure of glasses means that an exact, atom-for-atom registration across the fracture surface is necessary for healing, and this condition is extremely improbable. Even in a crystalline solid such as NaCl where crack healing has been observed (Forty and Forwood, 1963), the reknitting of the structure is quite imperfect because of disregistry.

It is worse than useless to use heat in an effort to heal microcracks, because each microcrack becomes a devitrification nucleus (Ernsberger, 1962, Fig. 11). The microcrystals formed have at least as bad an effect on strength as the original microcrack.

B. ETCHING WITH HYDROFLUORIC ACID

Concentrated hydrofluoric acid reagent (49% HF) reacts rapidly with soda-lime glass, with generation of noticeable heat, and deposition of insoluble CaF_2 and fluorosilicates. The resulting etched surface is very uneven. Dilution of the reagent by a factor of ten reduces the etching rate to about 1 μm/min, and the deposition of insoluble salts can be avoided entirely by agitation and frequent renewal of the solution. Under these conditions glass is removed uniformly from all exposed surfaces, including the surfaces of microcracks. As etching proceeds, therefore, the crack maintains its original depth, but steadily increases in radius of curvature. The strength of the glass increases rapidly at first and more slowly later on. The strength correlates quantitatively with the crack tip radius calculated from the removal (Proctor, 1964). Pavelchek and Doremus (1974) have suggested that the uniform-etching model of the etching process is oversimplified; that the rate of etching at the crack tip is much lower than that on the external surface. But the qualitative fact remains that any chemical reagent that can penetrate to the microcrack tip and enlarge the radius of curvature will strengthen the glass.

This procedure is a very valuable one for laboratory use. Not the least of its advantages is that the microcracks become visible etch pits, which makes it possible to study the distribution of strength-impairing damage sites on the surface. As a commercial technology, it is not very attractive. A major difficulty is that crystalline deposits begin to be formed before the HF content of the etch bath has been used up. Various ways to avoid this have been suggested, including vigorous air-bubble agitation and application of ultrasonics (Ryabov and Kupfer, 1970).

C. ETCHING WITH OTHER REAGENTS

Any reagent that removes glass uniformly from the surface will have a strengthening effect; even storage in cold water has a demonstrable effect (Mould, 1960). Certain chelating agents or mixtures thereof slowly dissolve glass (Ernsberger, 1959).

Steam decomposes soda-lime glass slowly, leaving a crust of hydrothermal reaction products. If, however, the glass is exposed to a high-velocity jet of steam containing entrained water droplets, the combination of corrosive and erosive effects results in a fairly uniform glass removal at a rate of about 20 μm/hr (Ryabov et al., 1972). Strength increases of fivefold to sixfold are obtained after the removal of about 200 μm of glass. Further increases can be produced but the glass loses transparency.

V. Nullification of Surface Damage

The strengthening techniques that are technologically successful are those that accept the presence of surface damage as unavoidable, but provide for the nullification of their effect. This can be accomplished by arranging to maintain the entire surface of the glass article in a state of compression. This negative stress bias must obviously be overcome before the surface microcracks can experience a potentially destructive positive stress.

A. PERIPHERAL CLAMPING

Surface compression in flat glass can be produced in a very straightforward way, simply by applying suitable forces along the periphery. For circular configurations, this can be done by shrinking a metal ring onto a glass disk; porthole windows have been made by this technique (Markov and Kopylov, 1970). For large glass areas and conventional rectangular shapes, this is obviously impractical.

B. CASING

In this and the following methods for surface-compression strengthening, the compressive moment at the surface is balanced by a tensile moment in the interior of the glass. Since the interior of glass is normally flawless, the existence of large interior tensions involves no risk of internal failure. There is, of course, a theoretical limitation because of the finite tensile strength of flawless glass, but such high internal stresses are never reached in practical cases. Shear stresses of substantial magnitude will occur at corners and edges of internally compensated glass articles, but again there is no practical concern because of the high strength of glass in shear. However, it goes without saying that cutting of internally stressed glass will cause shattering of the whole article if the internal tension is high enough.

In glass parlance, "casing" refers to the practice of overlaying one glass with another. Sometimes this is done for decorative effects, with glasses of like composition but differing colors. However, if the compositions are different, and the glass of lower thermal expansion coefficient is used for the outer layers, the cooling of such a glass "sandwich" will generate compression in the surface layers. There are many possible variations on this basic principle. No further elaboration will be undertaken here, because the state of this art is covered in another chapter.

C. THERMAL TEMPERING

It is unfortunate that the word "tempering" has been adopted to denote the technology of strengthening glass by heat treatment. The use of the term arose because of a superficial similarity to the process of tempering steel by heating and quenching. However, the mechanism involved and the effects produced are quite dissimilar in the two materials.

The theory and practice of thermal tempering are covered in detail in another chapter. For present purposes, it will be enough to define the process in terms of its essential elements, of which there are four.

First, the glass is heated to a uniform temperature at which it is fluid enough to relax internal stresses rather quickly, yet rigid enough to be handled without serious deformation; second, heat is rapidly and uniformly extracted from the entire surface so as to generate a symmetrical temperature profile across the glass thickness; third, the forced cooling is continued until the hottest point on the profile is below the effective solidification temperature of the glass. Finally, the glass is cooled to room temperature at any convenient rate. During this final phase of the cooling, the temperature profile relaxes toward uniformity. This relaxation leads to the development of internal stresses as the contraction of the relatively hot core is resisted by the cool surface layers.

The stress profile generated within the glass by this treatment is approximately parabolic. The balancing of moments for such a profile means that the zero-stress points on the profile are located approximately one-quarter of the glass thickness beneath each surface. The stress at the surface, where compression is a maxiumum, is about twice the stress in the central plane, where tension is a maximum.

This process is obviously simplest and most successful when applied to flat glass. Moderately bent shapes such as automotive glass are also processed successfully at the expense of complications in the engineeing. Hollow articles such as containers are seldom if ever thermally tempered because of the difficulty of cooling the interior.

The thickness of the article to be tempered is also a major consideration. As the glass becomes thinner, the heat transfer rate must be increased to maintain a fixed value of surface compression. Severe practical limitations are encountered at a thickness of about 3 mm with soda-lime glass. It is an obvious corollary that articles with substantial variations in thickness may not temper satisfactorily.

The composition of the glass has a definite influence on its response to thermal tempering, partly because of the composition dependence of the thermal expansion coefficient. The viscoelastic behavior of a glass in the transformation range also has a profound effect on its temperability. Other relevant properties that vary with composition are Young's modulus, Poisson's ratio, thermal conductivity, density, specific heat, and emissivity. However, the potential for influencing temperability through compositional variation of these properties is limited.

D. CHEMICAL STRENGTHENING

A discusion of the relatively new concept of "chemical" strengthening should begin with a definition. This is not easy, however, because a bewildering variety of treatments have come to be known as chemical methods for strengthening glass. These are fully covered in another chapter. The operative definition seems to be that any method for producing surface compression other than thermal quenching is regarded as chemical. Clearcut cases of chemical strengthening involve the use of chemical reagents to produce a surface region of altered chemical composition. In other cases, however, a thermal treatment with or without a chemical reagent may cause only a change in physical state, such as crystallization.

The common denominator of all the techniques that have come to be described as chemical is the fact that a way has been found to change the molar volume of the material in the vicinity of the surface and to do so in such a way that a compressive stress is produced. Molar volume in this connection is best defined as the volume per silicon atom, because silicon atoms are not added or removed in any of the known treatments.

The production of a compressive stress implies that the molar volume of the surface is *increased* by the treatment, but there are exceptions to this rule. In general, any change in physical or chemical state that is accompanied by a change in molar volume is a possible route to the generation of surface compression. Whether or not a significant degree of surface compression is realizable in practice depends to a large extent on the ingenuity with which available parameters such as composition, time, and temperature are manipulated.

An upper limit for the compressive stress that could in principle be generated from a given change in molar volume can be estimated from elasticity theory. Consider a hypothetical operation in which sufficient hydrostatic stress is applied to cancel the change in molar volume, followed by removal of one component of the hydrostatic stress. The two remaining stress components correspond to the two-dimensional hydrostatic state of stress in a compressed surface layer. The magnitude of either stress will be given by

$$\sigma = \epsilon E/(1 - \nu),$$

where ϵ is the edge deformation of a unit cube under the hydrostatic stress. E and ν are the elastic constants of the material. Thus for a 1% change in molar volume, the upper limit for the surface compression is about 44,000 psi. In this computation, typical values were used for E (10^7 psi) and v (0.25). Only a fraction of the theoretical stress can be achieved in practical cases because of viscoelastic relaxation and other effects.

E. MEASUREMENT OF SURFACE COMPRESSION

The useful strength of a surface-compressed glass is roughly equal to the sum of the annealed strength and the surface compression. Accordingly, it is desirable to be able to measure surface stress.

For thermally tempered glass it is usually sufficiently accurate to measure the maximum value of internal tension and multiply by two. Internal stress can be measured in terms of the relative retardation of e and o rays propagating in the central plane of the glass. The measurement is commonly done as a postmortem on one or more of the fragments remaining after tempered glass has been broken; however, it can be done nondestructively on a narrow test coupon. Direct observation of surface stress is relatively inaccurate because of the rapid rate of change of the stress profile near the surface.

Internal tension can be measured nondestructively on sheets of glass a foot or more in width, using light scattered by an intense polarized laser beam that traverses the central plane (Bateson *et al.*, 1966). By scanning the beam within that plane, the uniformity of internal stress can be assessed.

In chemically strengthened glasses, the stress profile is commonly a square wave rather than a parabola. Consequently, there is no simple invariant relationship between the magnitudes of surface compression and internal tension. Furthermore, the depth of surface compression may be so small that conventional compensation methods are useless for measuring retardation. Special instruments have been evolved for making surface-compression measurements in chemically strengthened glasses. These instruments are of two general types. Both measure the birefringence of the stressed surface, but they approach the measurement in different ways. The differential surface refractometer measures the small difference in the critical angle for total reflection for incident rays vibrating parallel and perpendicular to the surface, while the grazing-angle surface polarimeter or epibiascope measures the relative retardation of polarized rays propagating just beneath the surface and parallel with it. Both instruments require optical contact with the glass, which is secured by a wetted contact between the instrument optics and the glass surface. A paper by Ansevin (1965) should be consulted for details on the differential surface refractometer. The principles of the epibiascope are discussed by Guillemet and Acloque (1960).

VI. Serviceability of Strengthened Glass

The practical usefulness of a given technique for strengthening glass depends on the compatibility between the characteristics of the strengthened glass and the anticipated conditions of service. Some broad generalizations will be given in the following paragraphs.

A. Abrasion Effects

It has already been mentioned that the high strength of pristine glass is almost useless because of the weakening effect of even mildly abrasive service conditions. The same problem exists with some of the strengthening treatments of the surface-compression type, if the depth of the compression layer is too small. This deficiency is particularly likely in the case of chemical treatments. Most of these depend on the diffusional transport of reagents into the glass. Diffusion is an inherently slow process, and the penetration depth usually increases with only the one-half power of the time. The treatment temperature can be increased to speed diffusion, but undesirable effects supervene if the temperature is too high.

B. Time and Temperature Effects

At ordinary ambient temperatures and below, there is no observable rate of decay of the surface stresses produced by known methods of compression strengthening. However, certain applications of glass may require periodic or even constant exposure to elevated temperatures. Under these conditions, deleterious effects of two general kinds can occur.

One of these effects is an immediate loss of surface compression because of differential thermal expansion. Glasses strengthened by casing will of course show this effect, because differential contraction during cooling is the source of the stress. Certain types of chemically strengthened glass will also behave this way for the same reason. In all such cases the surface compression is restored upon cooling. The other kind of time–temperature effect is an irreversible decay of compression brought about by relaxation. The surface stress in thermally tempered glass begins to relax in this way as the annealing point is approached. An irreversible decay also occurs in glasses strengthened by diffusion. Where there is no external supply of the strengthening reagent, surface compression gradually diminishes at elevated temperatures through diffusive decay of the concentration gradient.

C. Static and Dynamic Fatigue

The highly covalent nature of the bonding in silicate glasses makes them immune to both static and dynamic fatigue, as long as the applied tensile stress does not exceed a threshold value. The exact value of this fatigue threshold is not well defined, but for unstrengthened glasses in the usual damaged condition, it cannot be more than about 2000 psi. This is too low to be of much use in load-bearing structures. However, glasses that are strengthened by surface compression have a fatigue threshold that may conservatively be taken as equal to the surface compression stress. This surface stress may be 50,000 psi or more. It follows that strengthened glass can be superior to metals, particularly for the design of cyclically stressed structures.

References

Ansevin, R. W. (1965). *ISA Trans.* **4**, 339–343.

Antonini, J. F., Hochstrasser, G., and Acloque, P. (1969). *Verres Ref.* **23**, 169–173.

Bateson, S., Dalby, D. A., and Sinha, N. K. (1966). *Bull. Am. Ceram. Soc.* **45**, 193–198.

Charles, R. J. (1958). *J. Appl. Phys.* **29**, 1549–1553.

Charles, R. J. (1966). U.S. Patent 3,275,470.

Ernsberger, F. M. (1959). *J. Am. Ceram. Soc.* **42**, 373–375.

Ernsberger, F. M. (1962). *In* "Advances in Glass Technology," Part 1, pp. 511–524. Plenum Press, New York.

Ernsberger, F. M. (1966). *Glass Ind.* **47**, 483–487, 542–545.

Ernsberger, F. M. (1967). U.S. Patent 3,345,148.

Ernsberger, F. M. (1969). *Phys. Chem. Glasses* **10**, 240–245.

Forty, A. J., and Forwood, C. T. (1963). *Trans. Brit. Ceram. Soc.* **62**, 715–724.

Griffith, A. A. (1921). *Philos. Trans. R. Soc.* **A221**, 163.

Guillemet, C., and Acloque, P. (1960). *In* "Compte Rendu du Colloque sur La Nature Des Surfaces Vitreuses Polies," pp. 121–134. Union Scientifique Continentale du Verre, Charleroi, Belgium.

LaCourse, W. C. (1972), *In Introd. Glass Sci. Proc. Tutorial Symp., 1970* (L. D. Pye, ed.), pp. 451–512. Plenum Press, New York.

Markov, V. P., and Kopylov, O. M. (1970). *Glass Ceram. (English Transl.)* **27**, 723–725.

Mould, R. E. (1960). *J. Am. Ceram. Soc.* **43**, 160–167.

Pavelchek, E. K., and Doremus, R. H. (1974). *J. Mater. Sci.* **9**, 1803–1808.

Proctor, B. A. (1964). *Appl. Mater. Res.* **3**, 28–34.

Ryabov, V. A., and Kupfer, V. V. (1970). *Glass Ind.* **51**, 268–271.

Ryabov, V. A., Semenov, N. I., and Paplauskas, A. B. (1972). *Glass Technol.* **13**, 168–170.

Scholes, A. B. (1971). *Proc. Int. Glass Congr., 9th*, Sect. A2.7, 1155–1166.

Shoemaker, A. F. (1969). *Mech. Eng.* **91**, 26–30.

Stookey, S. D. (1970). U.S. Patent 3,498,803.

Van Laethem, R. (1972). Belgium Patent 783,458. See *Chem. Abstr.* 9150n (1972).

Wiederhorn, S. M. (1969). *J. Am. Ceram. Soc.* **52**, 99–105.

Wiederhorn, S. M. (1970). *J. Am. Ceram. Soc.* **53**, 486–489.

CHAPTER 5

Thermal Tempering of Glass

Robert Gardon

ENGINEERING AND RESEARCH STAFF
FORD MOTOR COMPANY
DEARBORN, MICHIGAN

I. What Is Tempered Glass?

"Tempered glass" is so called by analogy to tempered steel. Both are strengthened by "tempering," a process that entails heating the material to a critical temperature and then rapidly quenching it. Here the analogy ends, for the immediate effects of this heat treatment are very different for the two materials. In steel, a new balance of hardness and toughness is produced by the precipitation of carbides. Glass, on the other hand, remains a single-phase material and its hardness is virtually unaffected.* Instead, tempering gives rise in glass to a stress system that raises its usable strength by permanently prestressing the surfaces in compression. It can do that because glass, like most brittle materials, is strong in compression but weak in tension. Since its failure almost invariably originates at a flaw in the surface, which acts as a stress multiplier for local tensile stresses, precompression of the surface allows a larger external load to be borne before the tensile strength is exceeded. In this respect tempered glass may be likened to prestressed concrete, though the nature and mechanism of the prestressing are, again, radically different.

A typical stress profile across the thickness of a tempered glass plate is shown in Fig. 1. The stress distribution is roughly parabolic, compression in the surface layers being balanced by tension in the interior. Since, there are usually no internal flaws in glass to act as stress raisers, this interior

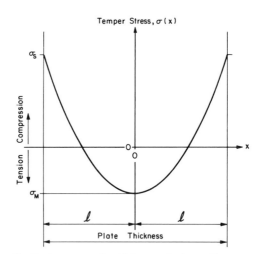

FIG. 1. Stress distribution across the thickness of a thermally tempered glass plate.

*The once current term "hardened glass" is therefore not appropriate. In Britain tempered glass is also called "toughened glass," which aptly describes its increased resistance to breakage.

tension is harmless. On the other hand, any externally imposed load that would tend to put the more vulnerable surface in tension now must neutralize the compressive prestress before any net tensile stress can begin to accumulate.

Seen from this viewpoint, any compression of the surface is useful. However, considering the wide spread of the strengths of flawed glass surfaces, the strengthening effect does not become significant until it exceeds the standard deviation of the breaking strength of the glass. Typically, the strength of annealed plate glass might be taken as 300 kg/cm² (4400 psi). The surface compression σ_S of commercial "fully tempered" glass is typically 1000 kg/cm² (14,700 psi), so that the effective strength of tempered glass might be 1300 kg/cm², corresponding to a fourfold to fivefold enhancement of strength.

This enhancement of breaking strength is not the only attribute sought in tempered glass. Another, in some applications equally important, is that, when breakage occurs, the fragments of tempered glass should be less dangerous than the sharp, often daggerlike fragments of annealed glass. Large fragments are a direct consequence of the low strength of annealed glass. Thus, when annealed glass breaks under a relatively low load, the strain energy in the glass at the moment of fracture is also low, and, correspondingly, little new surface is created in the fracture process. Badly flawed, weak glass may break into only two or three large pieces, having sharp edges but no points. In less badly flawed glass, breaking under a greater impact load, several cracks may propagate from the origin of the fracture, producing elongated fragments that are pointed as well as sharp. Not only is tempered glass stronger, so that more strain energy can be stored in it by greater external loads, but the temper stresses themselves impart to the glass a high content of strain energy even in the absence of external loading. Thus, when tempered glass is broken, its strain energy is always high enough to reduce the glass into usually harmless small fragments, more or less cubical in shape, with blunt (90°) edges and no sharp points (Fig. 2).

Unlike the strengthening effect, which we have seen to be more or less proportional to temper stresses, the "frozen-in" strain energy is proportional to the square of these stresses. Thus, while almost any degree of temper will have some strengthening effect, a reduction in the size of fragments becomes manifest only at higher levels of temper.

Several standards have been developed for commercially used tempered glass, notably for safety glazing of automobiles and buildings and for eye glasses. Written to reflect the best technology of its time, the U.S. automotive glazing standard of 1938 (ANSI, 1973) among other things stipulated a certain impact resistance and the weight (0.15 oz, ≈4.3 gm) of the largest permitted fragment in then current (¼ in., ≈0.6 mm thick) tempered glass.

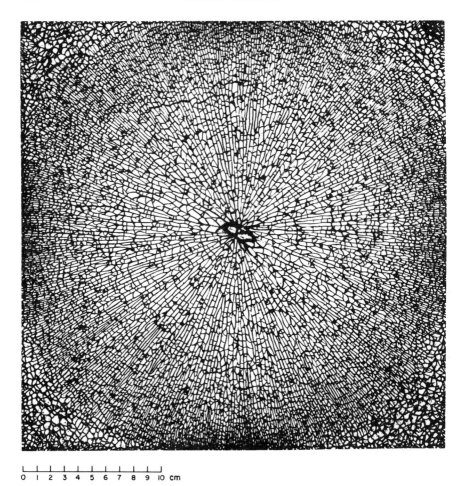

L I I I I I I I I I I J
0 I 2 3 4 5 6 7 8 9 10 cm

FIG. 2. Fracture pattern of a tempered glass plate broken by a sharp point. Note the two characteristically larger fragments at the fracture origin and the finer particles near the edges of the plate. Plate 30 cm square × 4 mm thick.

The degree of temper required to assure this has since been called, loosely, "full temper." It was usually measured, where such measurements were practicable (e.g., at the corners of large plates or on small specimens tempered along with the commercial product), by measuring the stress-induced birefringence in the midplane of the glass. For fully tempered glass this was, typically, 3200 nm/in. (1260 nm/cm), corresponding to a midplane tensile stress of approximately 6400 psi (435 kg/cm²), and a surface compressive stress of approximately 14,000 psi (960 kg/cm²). Semitempered glass, having a degree of temper roughly half of the above, is measurably stronger than annealed glass but breaks into fragments not unlike those of annealed glass.

The term fully tempered glass refers simply to compliance with certain commercial standards, representative of good industrial practice. It does not imply attainment of some upper limit of temper stresses that can be induced in glass by thermal tempering.

A. FRACTURE OF TEMPERED GLASS

The fracture of glass is discussed elsewhere in this volume. Here we shall briefly review only the special fracture characteristics of tempered glass. Noteworthy are the spontaneous propagation of a fracture, once initiated, the size of fragments produced, and the shape of the fracture front.

Tempered glass breaks in use when an external load causes tensile stresses in the glass surface to exceed the surface compression due to tempering by more than the tensile strength of the flawed surface. Alternatively, fracture may also be initiated—without supplying additional energy to the system—by introducing a flaw into the interior, where tempered glass is always under tension. This may be done by cutting into the glass with a diamond saw or by cracking the glass with a sharp, hard point. In either case, whether fracture is initiated externally or internally, its propagation is spontaneous, driven by the strain energy contained in the glass. It proceeds with frequent branching that leads to the creation of many small fragments. Roughly speaking, the larger the strain energy in the glass, the more new surface is created. Thus, when glass plates are broken in an identical manner and without addition of external energy, the average size of particles—or the "particle count" (per unit area)—is a rough measure of the degree of temper (Acloque, 1956; Barsom, 1968). It should be noted, though, that as Fig. 2 illustrates the size of fragments also depends on their position relative to the fracture origin and the edges of the plate.

From markings on fracture surfaces, the "Wallner lines," Acloque (1956) determined the shape of the fracture front in a tempered plate, as shown in Fig. 3. Evidently fractures in tempered glass propagate first in the interior that is in tension, and they penetrate to the surfaces only after compressive stresses there have been released as a result of the (local) release of tensions in the interior.

That fracture of the interior precedes fracture of the surfaces means that bifurcation of fractures is determined in the interior. This view is supported by ultrahigh-speed Schlieren cinematography (Fig. 4), which shows a network of faintly visible, internal fractures to precede appearance of the more sharply delineated fracture pattern that also becomes visible by ordinary illumination once fractures have penetrated to the surface (Acloque and Guillemet, 1962; Fayet et al., 1968).

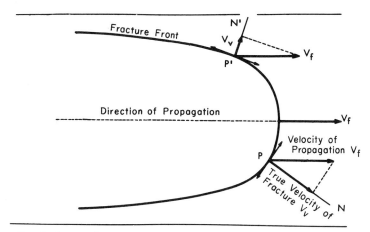

FIG. 3. Fracture front in a tempered glass plate (after Acloque and Guillemet, 1962).

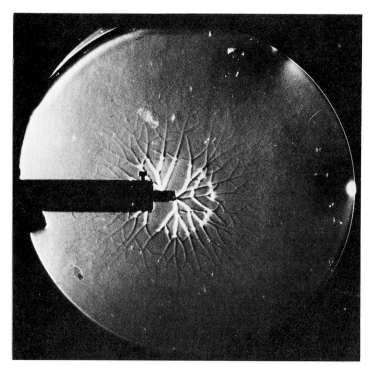

FIG. 4. Propagation of fractures in tempered glass broken by a sharp point (Fayet *et al.*, 1968).

Repeated bifurcation of fractures is brought about by two linked mechanisms:

(i)   A new crack will generally propagate in a direction normal to the largest local stress. As the crack grows, stresses normal to it are locally released, and as it penetrates farther into as yet unbroken glass, the ratio of affected volume to new fracture surface increases, making more strain energy available for crack propagation. The crack is therefore accelerated, and, as its speed increases, the central (tension) region of the fracture surface becomes progressively rougher.

(ii)   Simultaneously, as the unbroken regions on either side of a lengthening crack become larger, residual stresses at the crack tip and *parallel* to the crack also become larger. Thus there develops a growing tendency for a second fracture to open up at some angle to the first. After a while, the increasing roughness of the fracture surface at the apex of the fracture front provides a nucleus for a new crack to branch off the old one. Formation of this new crack close to the first one again reduces the strain energy locally

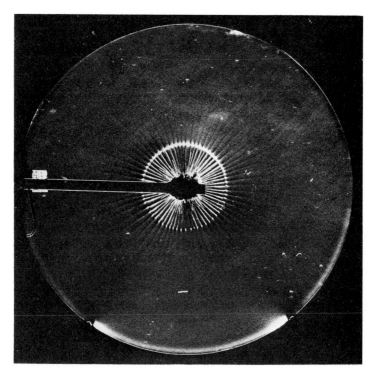

FIG. 5. Propagation of fractures in tempered glass broken by impact of a blunt object (Fayet *et al.*, 1968).

available to drive the cracks, so that, after bifurcation, they propagate with a lower velocity than the first crack had just before. Both cracks are thus set up for repetition of this process.

Residual stresses in the fragments thus formed and static fatigue of the glass give rise to further, delayed breakage. Secondary fractures are most marked in large, elongated fragments and are almost always at right angles to the primary fracture surfaces (cf. central region of Fig. 2).

One further point to be made here is that the fracture pattern illustrated in Figs. 2 and 4 is characteristic of ordinary tempered glass when broken with minimal addition of external energy, as for example by a sharp prick-punch. A very different fracture pattern can result if the stress system in the glass at the instant of fracture is significantly modified, as for example if the glass is broken by a blunt object that crushes the surface and also causes large bending stresses in the impacted plate. The resulting fracture pattern is illustrated by Fig. 5. Further discussion of this phenomenon is beyond our present scope, as are modifications of the fracture process in nonuniformly tempered glass (cf. Section V and Acloque, 1956).

II. Tempering and Tempered Glass

This section is a review of the salient experimental facts regarding stresses produced in glass and their dependence on operating conditions of the tempering process. All these experiments were performed on relatively small pieces of flat glass, since the measurement of temper stresses in objects of complex shape or even in large flat plates is rather difficult, while stresses in small parallelepipeds can readily be measured photoelastically and related to corresponding stresses in larger plates (Section VII.B.2).

A. NATURE OF TEMPER STRESSES

The stresses produced in a flat glass plate by rapidly chilling its surfaces are two-dimensionally isotropic plane stresses, constant over the extent of the plate (except for edge regions), and varying through its thickness only. A typical stress distribution is shown in Fig. 1. The stresses being internally balanced, photoelastic measurement is by far the most practical. As a result, temper stresses are frequently stated in units of stress-induced bire-fringence (S), such as nanometers (of retardation) per centimeter (of optical path length). Conversion to units of stress (σ) is via the stress-optical coefficient, which for plate or float glass has a value of about 0.39 (kg/cm²)/(nm/cm). Since the shape of stress distributions such as that shown in Fig. 1 is substantially constant, and since it is much easier to measure birefringence in the midplane than in the surface, where stress gradients are very steep, it has become customary to regard the midplane tension σ_M as a measure of the "degree of temper" (Bartenev, 1949). Actually, of course, it is the surface compression that enhances the strength of glass, and this was usually taken as from 2 to 2.2 times the midplane tension.* However, as Kitaigorodskii and Indenbom (1956) pointed out, the ratio of surface compression to midplane tension may also differ considerably from the above values, a circumstance that somewhat diminishes the utility of this definition of degree of temper (Section II.B and Figs. 8 and 9).

B. Dependence of Temper Stresses on Process Parameters

Figure 6 shows Bartenev's (1949) results in tempering glass by natural convection in air. The degree of temper produced depends strongly on both

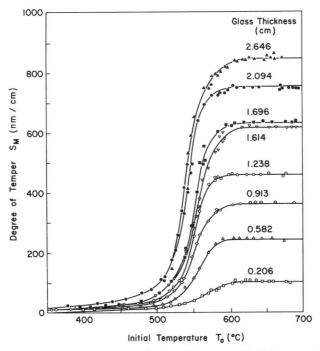

FIG. 6. Degree of temper as function of glass thickness and initial temperature. Natural convection cooling (Bartenev, 1949).

the thickness of the glass (L) and its initial temperature (T_0), as long as T_0 is below about 570°C. For higher values of T_0 the degree of temper attains a plateau, where it is independent of T_0. Even the highest "plateau levels" of temper attained in these experiments are quite modest: e.g., 250 nm/cm in 5.8 mm glass, i.e., only about $\frac{1}{5}$ of "full temper."

---

*It is a property of a balanced parabolic stress distribution of the form $y = ax^n$ that the ratio of surface compression $\sigma_s (= y_s - \bar{y})$ to midplane tension $\sigma_M (= \bar{y})$ *is equal to* n.

To attain higher temper stresses requires more intensive cooling by forced convection, and impinging air jets are commonly used to produce the required high heat transfer coefficients. A number of experimenters studied tempering with forced convection, among them Bartenev and Rozanova (1952), Kitaigorodskii and Indenbom (1956), and Acloque (1961). The heat transfer coefficients involved were first measured independently of tempering experiments by Gardon and co-workers (1961, 1966) and used in subsequent studies to determine their effect on the degree of temper produced (Gardon, 1965a). The solid curves in Fig. 7 illustrate some of the

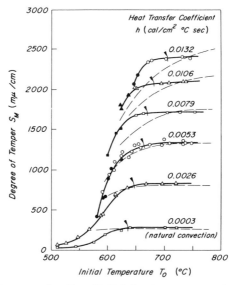

FIG. 7. Degree of temper as function of initial glass temperature and quenching coefficient. Forced convection cooling. Glass thickness, 0.61 cm. Experimental results, —○—; broken specimens, ●; calculated by Bartenev's theory ––– (Gardon, 1965a).

results. They are similar to Bartenev's (Fig. 6) in that they show the degree of temper to attain a plateau and become independent of T_0 above 625–675°C. Also, the lowest curve in Fig. 7—that for $h = 0.0003$ cal/cm² °C sec, which corresponds to cooling by natural convection and radiation—closely follows the curve in Fig. 6 for glass of nearly the same thickness. However, Figs. 6 and 7 differ in that, while T_0 can be reduced to 400°C when glass of any thickness is quenched by natural convection, this is not the case for any but the lowest heat transfer coefficients produced by forced convection. With the latter, quenching from too low an initial temperature results in fracture of the glass during tempering. Most such fractures appear to be caused by transient stresses due to preferential quenching of edges of the specimen.* The resulting (usually) single crack does not diminish the degree of temper produced in the fragments. This is clearly shown in Fig. 7 by the fact that results obtained for $h = 0.053$ cal/cm² °C sec with specimens that broke (solid symbols) and those that did not break (open symbols) overlap and define the same curve of S_M versus T_0.

It appears that, at low cooling rates, the initial temperature can be reduced indefinitely without risking fracture. All that happens is that the degree of temper is also reduced once the low-temperature endpoints of the plateaus have been passed. At a moderate quenching rate of $h = 0.005$ cal/cm² °C sec, which is representative of industrial practice for tempering 6-mm glass, a reduction of the initial temperature below the end of the plateau leads not only to less strongly tempered glass but also to a progressively greater probability of fracture. For the highest quenching rates of these experiments, the low-temperature endpoints of the plateaus also represent the lowest temperatures from which specimens could be tempered without breakage.

One practical implication of these results is that it is clearly advantageous to operate the tempering process with an initial temperature near the low limit of the plateau: this will hold down viscous flow and distortion of the glass and yet allow attainment of the highest degree of temper possible with a given quenching rate. All too often, however, the desire to maintain good geometrical form of the glass, even only flatness of large plates, prompts operators to lower initial temperatures yet further. Under some conditions, the resulting reduction in temper can be partially offset by going to somewhat higher quenching rates, although this is likely to cause some in-process breakage. At high quenching rates, there is little latitude between a T_0 high enough to achieve the desired high degree of temper without excessive breakage and a T_0 low enough to maintain good form.

If, in Fig. 7, we had plotted surface compression rather than midplane tension, substantially similar curves would have been obtained, only that surface stresses obtainable by rapid quenching continue to increase with increasing T_0 even after midplane stresses have attained their plateau levels. Figure 8 presents these results for plate glass 0.6 cm thick, expressed in more general terms by showing the ratio of surface compression to midplane tension as a function of T_0 and h. For quenching from sufficiently high initial temperatures, this ratio is seen also to attain plateau levels, which for a given thickness of glass depend only on the heat transfer coefficient h. Figure 9 (from Boguslavskii et al., 1964) shows plateau levels of

---

*These stresses are not intrinsic to tempering and will be discussed in Section V. However, even in their absence, glass quenched from too low a T_0 would tend to break from excessive transient tensile stresses in the surface (cf. Section II.C).

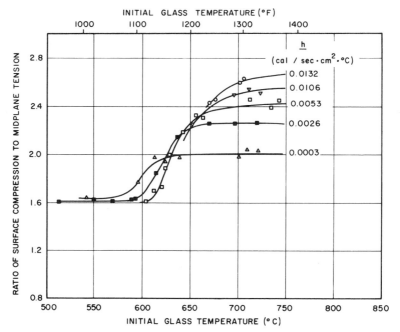

FIG. 8. Ratio of surface compression to midplane tension as function of initial glass temperature and quenching coefficient. Glass thickness $L = 0.61$ cm; coolant temperature $T_a = 25°C$. (Gardon, 1965b).

this ratio as functions of the Biot number, Bi $= hl/k$, thus also covering results obtained with glass of different thicknesses $2l$. These experiments are noteworthy for encompassing extremely high heat transfer coefficients, such as are obtainable by quenching in liquids. While the ratio of surface compression to midplane tension obtainable with air quenching rarely goes

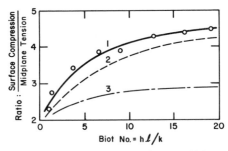

FIG. 9. Ratio of "plateau level" surface compression to midplane tension as function of Biot number. Curve 1, experimental results; 2, predictions by Indenbom's theory; 3, predictions by Bartenev's theory (Boguslavskii *et al.*, 1964).

above 2.5, values in excess of four have been obtained with liquid quenching media.

C. Transient Stresses During Tempering

Transient stresses during quenching were first studied by Acloque (1950, 1961), who used cinematography to record the corresponding transient photoelastic fringe patterns. Figure 10 (from Gardon, 1965a) illustrates the

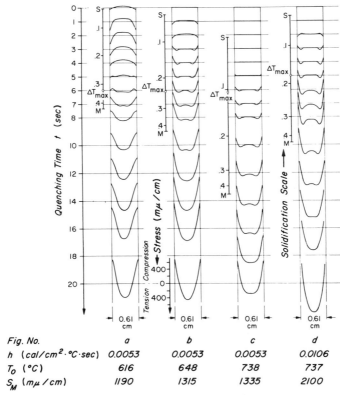

Fig. No.	a	b	c	d
h $(cal/cm^2 \cdot °C \cdot sec)$	0.0053	0.0053	0.0053	0.0106
T_0 $(°C)$	616	648	738	737
S_M $(m\mu/cm)$	1190	1315	1335	2100

FIG. 10. Transient stress distributions in glass plates during quenching (after Gardon, 1965a).

development of temper stresses under four characteristically different operating conditions. These four sequences, also obtained cinematographically, show stress profiles in the glass at intervals of 1 or 2 sec for the first 20 sec of quenching. The final degree of temper, in terms of the midplane tension S_M, is stated at the bottom of each sequence, along with the operating conditions used to produce it. The figure shows that transient stresses in the glass during the first 20 sec of quenching are quite small in comparison with the final temper stresses. Indeed, the characteristic parabolic stress distributions do not begin to develop until relatively late in the quenching process, after the glass has passed through the transformation range and as temperature differences within the glass decay. Stress distributions during the early stages of quenching exhibit a variety of very different shapes. Some early stress profiles in initially relatively cold glass (Fig. 10a) are more or less parabolic, though of opposite sign to the final stresses. In glass quenched from high initial temperatures (cf. Figs. 10c and d), stresses early in the process are clearly confined to the (colder) surface regions, with the interior remaining totally stress-free for several seconds. Note also that early stresses in the surface may be tensile or compressive, depending on the initial temperature T_0. If this is low (cases a and b), the surfaces go into tension before they go into compression, while in higher temperature runs (c and d) they pass directly from an initially stress-free condition into compression. Figure 10 will be more fully discussed in Sections III and IV in the context of developing and evaluating various theories of tempering.

D. INFLUENCE OF TEMPERING ON PHYSICAL PROPERTIES OF GLASS

For all the marked effect of temper stresses on the usable strength of glass, tempering has very little effect on its other physical properties. One consequence of this is that past searches for an alternative to photoelasticity as a nondestructive means to measure temper have been fruitless: The visible and infrared transmission of glass is unaffected by tempering. Young's modulus is reduced by about 3% (Kerper and Scuderi, 1966). Changes in density (of the order of 0.1%) and refractive index are also too small to serve as practical means to measure temper, or even to confirm that glass has been tempered. However, the latter changes are interesting as means to characterize structural changes in glass that accompany quenching and, in turn, also affect the generation of temper stresses (Sections III.B4 and IV.E).

Some representative density distributions in tempered glass are shown in Fig. 11 (from Gardon, 1978). Of the five specimens of the same glass, one was annealed and four were quenched at different rates from a fixed initial temperature of 677°C. The left-hand side of the figure shows the density distributions (ρ_1) in the specimens as tempered, the right the stress-free density distributions (ρ_3) in small fragments in which all temper stresses have been released. It may be seen that the average densities of the samples decrease with increasing severity of quenching. Superimposed on this variation of average densities are spatial density distributions that vary characteristically between the whole specimens and their fragments. In the glass as tempered, densities (ρ_1) are higher at the surface than in the inte-

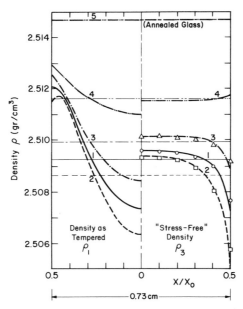

FIG. 11. Density distributions in tempered glass. Density distributions shown by heavy curves; average densities by thin horizontal lines. Numbers identify specimens, in order of increasing temper: 5,4,3,1,2. No. 1 represents commercial "full temper" (1408 nm/cm midplane tension) (Gardon, 1978).

rior. This is due to compression of the surface layers and dilatation of the interior by temper stresses. The interesting point to note is that these stresses are impressed upon an inhomogeneous material, the true, i.e., stress-free, density (ρ_3) of which varies in the opposite manner, the surfaces being less dense than the interior. As we shall see, this results from the different effective cooling rates of different layers of glass, as a consequence of which the surfaces "set" at a higher "fictive temperature" than the interior (Narayanaswamy, 1978).

III. Physics of the Thermal Tempering Process

In a review of theories of annealing and tempering, Acloque (1951) noted that these two processes are sometimes considered together for no better reason than that, when indifferently conducted, they produce glass that can equally be called "poorly annealed" or "poorly tempered." Some parallels clearly exist between annealing and tempering, both as to the mechanism of stress generation and as to the shape, even if not the magnitude, of the resulting stress distributions. To the extent that the two processes have common elements, a discussion of tempering may well lean on Adams and

Williamson's (1920) classic treatment of annealing. As we shall see, however, the two processes also differ in several significant respects. As a result, some physical mechanisms enter into tempering that are not involved in annealing.

The two processes differ, right from the start, as to the heat treatments employed: Annealing entails slow cooling of the glass in a furnace or annealing lehr, the temperature of which is controlled, typically to produce a constant cooling rate through the annealing range. Tempering, by contrast, entails quenching of the hot glass into a cold medium at constant temperature, so that the cooling rate of the glass is both high and necessarily varying with time. A typical temperature—time history is illustrated in Fig. 12a.

In the first phase of quenching, the surfaces of the glass solidify and contract while its core is still relatively fluid. By the time the core has also set, its contraction is resisted by the already solid surface layers, which are thus put into compression, while the core itself is put in tension. Temper stresses may thus be regarded as thermal stresses that remain in the glass even in the absence of the transient temperature differences that evoked them. For this freezing in of stresses to occur, the rapid change of viscosity with temperature that is characteristic of the glass transition is a prerequisite.

A. The Simplest View: Thermoelastic and Permanent Stresses in Glass

For simplicity, and partly also because tempering is most widely used in the flat glass industry, we shall consider the tempering of a flat plate. The physical mechanisms involved are, of course, equally applicable to other shapes.

Figure 12a schematically represents the temperature–time history of the surface and midplane of a glass plate, initially at a uniform temperature T_0, that is rapidly quenched on both surfaces. The quenching medium is taken to be at room temperature, and the coefficient of heat transfer as constant. Figure 12b shows the variation of the temperature difference ΔT between the midplane and surface. Note that this temperature difference at first increases rapidly, reaches a maximum, and then decays to zero as the glass returns to isothermal conditions at room temperature.

If we were dealing with a purely elastic material (or, in the case of glass, if T_0 were well below the strain point), thermal stresses in the plate would parallel the temperature differences causing them: colder regions of the glass would tend to contract, but be partially restrained by warmer regions, so that regions below the average temperature of the plate would be in tension, those above in compression. Except for a brief transient immediately following the onset of quenching, temperature distributions in the

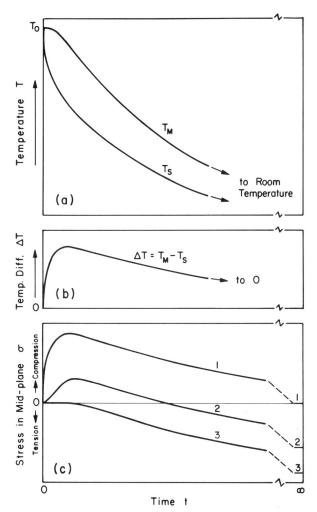

FIG. 12. Schematic temperature- and stress-histories during the tempering of a glass plate. (c): Curve 1, $T_0 \ll$ strain point (elastic plate); 2, $T_0 \geqslant$ anneal point (too low for tempering); 3, $T_0 \gg$ anneal point (about right for tempering).

plate are substantially parabolic. It follows that stresses in the plate also have a parabolic distribution, with tensile stresses in the colder surface layers being balanced by compression in the interior. Stresses would increase while ΔT increases and decrease as ΔT decreases. They would vanish when, at ambient conditions, ΔT shall have vanished. Indeed, the plot of ΔT versus time—to an appropriate scale—may also be regarded as the stress–history of the midplane of an elastic plate, as indicated by curve 1 of Fig. 12c.

Note that in this elastic plate no residual or permanent stresses are produced as a result of quenching. Also, transient stresses in the surface are tensile. This explains the well-known fact that glass is liable to break when quenched from too low an initial temperature.

The likelihood of breakage is reduced and some permanent stresses are created in the glass if its initial temperature is raised to a level at which stress relaxation by viscous flow can come into play. As quenching begins, the surfaces tend, as in the elastic case, to go into tension, the midplane into compression. Now, however, this is partly offset by a simultaneous process of stress relaxation, so that stresses rise less rapidly than in the elastic material. The maximum transient compression in the midplane is, correspondingly, also lower, as shown by curve 2 in Fig. 12c. Stresses in this initially heat-softened material are thus seen to be lower than they would be in an otherwise identical but always elastic material having the same temperature distribution. Once the glass has cooled to below the transformation range and all further stress relaxation has stopped, it becomes elastic again. From here on, therefore, further changes in stresses are determined solely by changes in the temperature distribution, so that curve 2 for the glass plate runs parallel to curve 1 for the elastic plate. In a material that had been elastic throughout this process, the stresses produced by the decay of temperature gradients would be equal and opposite to those produced by their growth. In case 2, these two are not equal, since some of the stresses generated during the early phase of cooling were relaxed by viscous flow. As a result, the glass ends up with some stresses in it even after its temperature has become uniform. These are the "frozen-in" temper stresses.

If the initial temperature T_0 is raised higher still, the initial rate of stress relaxation may be high enough that no stresses become manifest (in the midplane, at least) until the glass has cooled for a few moments. If "solidification" is thus delayed till after the maximum temperature difference ΔT_{max} has been passed, then all temporary compression of the midplane is avoided, as shown by curve 3. Tensile stresses in the surface are also largely, if not completely, avoided. (For a closer look at surface stresses in this case, see Section III.B.3.) This condition thus leads to a further reduction of the risk of breakage during tempering and also to the highest attainable permanent stress for the given quenching rate.

This argument has already been carried further than the underlying simplifications warrant, and we shall return to refine it in Section III.B. It was presented here to make three basic points:

(i) *Permanent stresses* arise in tempered glass primarily as a result of the relaxation of thermoelastic stresses early during quenching, and, to a first approximation, they are equal in magnitude and opposite in sign to the thermoelastic stresses relaxed [cf. reference to Adams and Williamson (1920) in Section III.B.2.].

(ii) *Transient stresses* that appear in the course of cooling can also be important, not least because they can, under adverse conditions, result in breakage of the glass. Transient stresses appear when the rate of stress relaxation, which decreases with decreasing temperature, can no longer keep up with the rate at which stresses tend to be generated by changing temperature distributions. Thus early appearance of stresses is promoted by lower initial temperatures, as we have seen from Fig. 12c, and also by higher cooling rates.

(iii) The sign of stresses tending to be accumulated depends on whether temperature differences in the glass (such as the ΔT shown in Fig. 12b) are increasing or decreasing.

All these points will be considered further in Sections III.B and IV.

B. REFINEMENTS: THERMAL HISTORY, TEMPERATURE EQUALIZATION, STRESSES, SOLIDIFICATION STRESSES, AND STRUCTURAL EFFECTS

One shortcoming of Fig. 12c and its discussion is that they focused on transient stresses in the midplane only. In an elastic material, where the *shape* of the stress distribution is more or less fixed, the midplane stress also serves as a measure of stresses elsewhere. This also holds, approximately at least, for glass quenched from temperatures not much above the transformation range, as may be seen by comparing curve 2 of Fig. 12c with the sequence of midplane stresses shown in Fig. 10a. It clearly does not hold for glass quenched from higher temperatures, where transient stresses may arise in the surface that are in no way reflected by stresses in the midplane, as witness Fig. 10d. Curve 3 of Fig. 12c thus shows stresses in the midplane that are not representative of stresses in other parts of the glass.

1. Thermal History

Refinement of the preceding simple view of tempering might well begin with a closer look at the thermal history of a glass plate as it undergoes tempering. While in Fig. 12b we considered the temperature difference between the surface and midplane only, Fig. 13 shows the variation with time of temperature differences across "slices" of glass of thickness $l/5$, denoted by $\Delta T_1, \Delta T_2, \ldots, \Delta T_5$. The overall ΔT shown in both Figs. 12b and 13 is clearly the sum of the above. The point to be made here is that, while the midplane will tend to go into compression as long as the overall ΔT is increasing, the tendency for the surface to go into tension ceases earlier. The reason for this is that, while the surface (plane 5) continues to

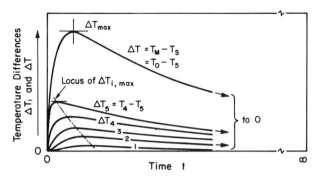

FIG. 13. Local temperature differences in glass as a function of position and time. ΔT_i denotes temperature difference between planes located at $(i - 1)l/5$ and $il/5$ from the midplane. Subscript 0 refers to midplane, 5 to surface.

contract relative to the midplane as long as ΔT is increasing, it starts to expand relative to plane 4 ($l/5$ below the surface) as soon as ΔT_5 has passed its maximum. As Fig. 13 shows, this occurs before ΔT_{max} is reached. Thus, to avoid transient tension in the surface, its solidification need not be delayed quite as long as suggested by Fig. 12c.

Yet more precisely than Fig. 13 suggests, the *tendency* of any layer of a cooled plate to go into tension or compression is governed by whether local temperature gradients are increasing or decreasing, i.e., by the sign of $\partial^2 T/\partial x\,\partial t$.

2. Temperature Equalization Stresses

To refine further the simple view of tempering presented in Section III.A, it must be recognized that "solidification" of the glass, as it is being cooled, does not occur all at once. First, the transformation range itself extends over a temperature interval of 50–100°C, depending on the rate of cooling. Second, different layers of glass pass through any given temperature at different times. At the quenching rates customarily used in tempering, temperature differences between the midplane and surface of 100°C are not uncommon. Thus "solidification" of a glass plate starts when the surface passes the upper limit of the transformation range and ends when the midplane passes the lower limit. It is only from this point on that the glass is wholly elastic, so that further changes in stress distribution are uniquely determined by further changes in temperature distribution. We shall refer to the corresponding component of temper stresses as "temperature equalization stresses." These are the stresses evoked in the now wholly elastic glass by the decay of the temperature distribution existing in it as it finally emerges from the transformation range.

In annealing, glass is cooled through the annealing range at a more-or-less constant rate and with a correspondingly invariant temperature distribution through its thickness. If such cooling starts from a temperature high enough for the glass to be initially stress-free, the glass will remain stress-free through the annealing range and even below the strain point, since the mere fact of "solidification" only makes the existence of stresses possible, but does not actually bring them about. Only as the cooling ends and isothermal conditions are reestablished at room temperature does the decay of the temperature distribution that existed during cooling bring forth corresponding thermoelastic stresses.* We recognize them as temperature equalization stresses. It was for these that Adams and Williamson (1920) enunciated their well-known generalization to the effect that "the strain remaining in glass is equal and opposite in sign to the . . . strain lost by viscous yielding in the early stages of the cooling process." Approximately, at least, we have seen this also to hold for tempering. But, while temperature equalization stresses are the only stresses brought about by annealing, this is not the case for tempering.

3. Solidification Stresses

Tempering, unlike annealing, is specified not by a cooling rate but by the temperature of the quenching medium and the applicable heat transfer coefficient. The cooling rates involved are orders of magnitude higher than in annealing and must necessarily also vary with time: rates of 10–100°C/sec simply cannot remain constant very long. Temperature distributions change correspondingly, even during the brief interval in which parts of the glass are within the transformation range. As a result, thermoelastic stresses are evoked while the glass, or parts of it, are still within the transformation range. While stress relaxation may reduce them, or even nullify them, in annealing by cooling at a constant rate none would have been evoked to begin with. Whatever stresses generated in the transformation range are not relaxed by the time the glass emerges from the transformation range constitute a distinct additional component to the final permanent stresses in the glass. We shall refer to them as "solidification stresses," a term first used, in a somewhat different context, by Kalman (1959).

Two mechanisms must be recognized as factors determining the magnitude and distribution of solidification stresses.

In the first place, and locally, the appearance of stresses while the glass is still in the transformation range depends on an imbalance between local

*This is the classical view of annealing, as put forward by Adams and Williamson. For a different view, see Gardon and Narayanaswamy (1970). They showed that stresses of structural origin arise even while glass is being cooled at a constant rate. The theoretical treatment of this phenomenon was a forerunner of the "structural" theory of tempering discussed in Section IV.E.

rates of stress generation and local rates of stress relaxation. The rate of stress generation depends on the rate of change of temperature gradients, which is initially very high at the surfaces. The rate of stress relaxation decreases rapidly with decreasing temperature, and therefore most rapidly at the surfaces. Both factors favor the early appearance of stresses in regions near the surfaces. In some cases, these stresses appear and are balanced within the surface region, while the midplane is still totally stress-free (Figs. 10c and d). The further evolution of stresses continues to depend on the imbalance of rates of stress generation and relaxation. Stress relaxation always makes for a reduction of existing stresses, though at a diminishing rate as the temperature of the glass falls and its viscosity increases. The rate of stress generation may change sign as (local) temperature gradients in the glass pass their maxima. (See Fig. 13, and Acloque, 1961).

In the second place, and looking at the glass plate as a whole, it is clear that stress relaxation by viscous flow in any part of the glass still hot enough for this to occur will also affect stresses in colder parts of the glass, since stresses through the thickness of the glass must balance. Solidification stresses are thus determined not only by the imbalance of local rates of stress generation and relaxation, but also by a redistribution of stresses within already elastic regions of the glass, brought about by continuing relaxation in hotter regions. Clearly, Adams and Williamson's generalization does not apply to this latter mechanism.

4. Structural Effects

One last distinction in the nature of stresses needs to be made. All the stresses so far considered, whether they are fully or only partially "remembered" by the glass, are thermoelastic in origin, i.e., they are caused by changes in length with temperature.

In tempering, the surfaces of a glass plate are usually cooled through the transformation range more rapidly than the midplane. This gives rise to a heterogeneity of the structure of the glass, which among other things also manifests itself by a variation of its (stress-free) density with position, as shown in Fig. 11. Acloque (1951) raised the question whether density differences of structural origin also give rise to stresses, and how these interact with stresses of thermoelastic origin. In the latest stages of research on tempering, this question has now also been answered (Narayanaswamy, 1978), so that we may distinguish between solidification stresses of "mechanical" or thermoelastic origin and those of "structural" origin (cf. Section IV.E).

With this outline of the physical mechanisms involved in tempering, the stage has been set for a brief consideration of the various theories put forward to describe tempering quantitatively.

IV. Theories of Tempering

The preceding section was an entirely qualitative overview of the physical mechanisms involved in tempering. The following more-or-less chronological review of quantitative treatments of tempering seeks to follow the same outline, proceeding from temperature equalization stresses to progressively more complex treatments of tempering that also take account of solidification stresses. Distinctions between the effects of various physical mechanisms discussed in the preceding section have often been lost in the mathematical complexity of some of the theories. In the following, an attempt is made to redress this by outlining mathematical treatments in the context of their physical meaning and with a focus on the new contributions various authors have made to the understanding of tempering. It is hoped that this approach may make the mathematical treatment of the subject more readily comprehensible to nonspecialists and help to establish a clearer overview of the subject and its historical development.

Certain elements are common to all theories of tempering, and these are briefly reviewed first, so that we may later concentrate on the new elements of succeeding theories.

A. General Considerations

All of the following theories of tempering were developed for a flat plate that is large in relation to its thickness and is uniformly and symmetrically cooled on its two faces. Considerations are restricted to a region away from all edges. The coordinate system is centered on the midplane of the plate (Fig. 14).

Under these conditions temper stresses are two-dimensionally isotropic, uniform plane stresses in y–z planes that vary only with x. In view of symmetry, x, y, and z are principal axes. Thus

$$\sigma_y = \sigma_z = \sigma = \sigma(x), \tag{1a}$$

and

$$\sigma_x = 0. \tag{1b}$$

Any theoretical treatment of tempering must start with means to predict the temperature history of the glass as it is being quenched. Such a temperature history is schematically illustrated in Figs. 12 and 13. The heat treatment involved is specified by the initial temperature T_0 of the glass, the temperature T_a of the quenching medium (usually air), and the heat transfer coefficient h governing the rate of heat exchange between the two. Once these are known, the calculation of glass temperatures is an exercise in transient thermal conduction, thermal effects of viscous flow in the glass

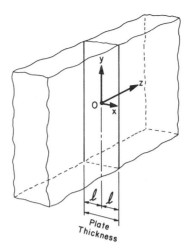

FIG. 14. Coordinate system.

being wholly negligible. The cases of interest here are well covered by Car-slaw and Jaeger's (1959) text, which gives exact solutions for temperature as a function of x and t in the form of infinite series. With the advent of high-speed computers, temperatures can also be calculated by various numerical techniques. One of these (Gardon, 1958) also makes allowance for radiative cooling, but—at the temperatures involved in tempering—this is not significant for glass that is rapidly quenched. For glass that is rela-tively slowly cooled, e.g., by natural convection, radiative cooling can be significant, although the temperature gradients it produces in a partly trans-parent material are smaller than they would be in an otherwise similar, but opaque material. For clear glass quenched from about 650°C, radiative effects in tempering can be allowed for by augmenting the convective heat transfer coefficient by an effective radiative heat transfer coefficient of h_{rad} = 0.0007 cal/cm² °C sec (Gardon, 1965a).

Once the temperature–position–time history of the glass is known, its mechanical response to heat treatment is governed by three stress-analyti-cal considerations:

(i) The condition of (geometrical) *compatibility* of strains. Since the plate must expand or contract as a whole, this requires that strains be inde-pendent of x. They can therefore vary only with time. Thus

$$\epsilon_y = \epsilon_z = \epsilon = \epsilon(t). \tag{2}$$

(ii) The *equilibrium* of the plate requires that stresses be balanced, i.e., that the average stress through the thickness be zero. Thus

$$\overline{\sigma(x,t)} = \frac{1}{l} \int_0^l \sigma(x,t) \, dx = 0. \tag{3}$$

(iii)   The law governing stress relaxation, which is also the *stress–strain relation* for the glass. Various theories of tempering differ primarily in the assumptions made regarding this point. The theories have been grouped accordingly, and they are discussed in Sections IV.C–E, together with the underlying stress–strain relations.

B. Temperature Equalization Stresses

These are the stresses evoked by the decay of the temperature distribution that exists in a cooled glass plate as it finally emerges from the transformation range. Since the glass is then entirely elastic, temperature equalization stresses can readily be calculated from first principles of thermoelasticity (Timoshenko and Goodier, 1970, p. 433).

If $T(x)$ represents the instantaneous temperature distribution across the thickness of a glass plate at the instant its midplane (i.e., its hottest part) becomes elastic, then

$$\sigma_{\text{TE}}(x) = [E\beta/(1 - \nu)] [\overline{T(x)} - T(x)], \tag{4}$$

where σ_{TE} is the temperature equalization stress, $\overline{T(x)}$ the average temperature corresponding to the temperature distribution $T(x)$, E is Young's modulus, ν is Poisson's ratio, and β the coefficient of linear expansion. The average temperature $\overline{T(x)}$ enters into Eq. (4) to satisfy the equilibrium condition [Eq. (3)], i.e., that forces across any section of the plate be balanced.

Stresses developed by cooling glass through the annealing range at a constant rate are a special case of temperature equalization stresses, of interest in annealing and discussed by Adams and Williamson (1920). During cooling at a constant rate, a parabolic and time-invariant temperature distribution is established across the thickness of the plate, given by

$$\overline{T(x)} - T(x) = \frac{Rl^2\,c\rho}{2k} \left(\frac{1}{3} - \frac{x^2}{l^2} \right), \tag{5}$$

where l is the half-thickness of plate, R the cooling rate (positive for cooling), $c\rho$ the volumetric specific heat, and k the thermal conductivity. The stress distribution resulting from the decay of the above temperature distribution is also parabolic, with the surfaces of the glass under compression and the interior under tension. With the *shape* of the stress distribution thus fixed, it is convenient to use the midplane tension as a measure of the permanent stresses frozen into the glass. This stress is given by

$$\sigma_{\text{M}} = \sigma_{\text{TE}}(0) = \frac{E\beta}{1 - \nu} \frac{R\,l^2 c\rho}{6k}, \tag{6}$$

a useful relation for annealing, which is usually specified by its cooling rate. Matters are different with tempering, in which cooling rates vary markedly with both time and position in the plate. This not only renders Eq. (6) inapplicable, but also means that the temperature equalization stresses given by Eq. (4) are *not* the only stresses produced by tempering. They are, however, the dominant stresses in many cases, and they were the only ones considered in early theories of tempering (e.g., Reis, 1933).

C. "Instant Freezing" Theories

To take account of solidification stresses in a quantitative manner requires a model of stress relaxation in a form simple enough to be used in conjunction with an analysis of the temperature–time history of the glass and of the stresses produced by this. The earliest attempt at this was that of Bartenev (1948, 1949), who developed the first of what we now call "instant freezing" theories.

Bartenev (1948a) postulated that—in view of the rapid variation of the viscosity of glass with temperature and of temperature with time during tempering—glass be treated as "solidifying" at a single "solidification temperature" T_g. In this simple model, glass above T_g is treated as a fluid incapable of supporting any stress, while glass below T_g is regarded as an elastic solid in which no flow can occur and no stress relaxation. Accordingly, the solidification of a glass plate progresses simply as the T_g isotherm travels from the surface to the midplane, as indicated in Fig. 15.

On this basis, Bartenev (1949) treated the permanent stresses produced by quenching as resulting from the decay, in an initially stress-free, elastic medium, of a fictitious temperature distribution $\Phi(x)$, to be obtained from

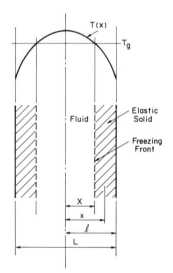

Fig. 15. The "instant freezing" model.

the actual temperature gradients in successive layers of glass at the instants of their solidification. Thus he defined $\Phi(x)$ by the equation

$$\frac{d\Phi}{dx} = \frac{dT}{dx}\bigg|_{T = T_g} , \tag{7}$$

and, analogous to Eq. (4), he obtained the stress at any position x within the plate as

$$\sigma(x) = [E\beta/(1 - v)]\,[\overline{\Phi(x)} - \Phi(x)]. \tag{8}$$

Note that, while Bartenev did not explicitly distinguish between stresses brought forth during the period of solidification and the subsequent period of temperature equalization, his definition of Φ represents an attempt to take account of both as far as final temper stresses are concerned. Thus, to the extent that $\Phi(x)$ differs from the temperature distribution $T(x)$ at the instant the midplane passes through T_g, that difference represents a measure of what we have called "solidification stresses." However, inconsistencies in Bartenev's mathematical treatment do not allow meaningful transient stresses to be derived from it. We shall return to this point in the context of Indenbom's treatment of the "instant freezing" hypothesis.

Bartenev proceeded to develop this approach specifically for what he called "regular" conditions of tempering, in which solidification occurs only after the temperature difference between the midplane and surface of the plate has passed its maximum (Fig. 12b). This allowed him to simplify the calculation of temperatures in the glass and is also desirable for the practical reason that it largely avoids temporary tensile stresses in the surface (Section II.C). For this condition, which of course implies a relatively high initial temperature T_0, he found $\Phi(x)$ and therefore also the degree of temper to be independent of T_0.

Bartenev verified his theoretical treatment experimentally for a range of temper stresses obtained by varying the thickness of his glass specimens (Fig. 6). The level of temper stresses attained in these experiments was quite low, however, since only cooling by natural convection was used. For these conditions the predictions of his theory were found to be very good, the more so since the heat transfer coefficient and the values of T_g involved were determined from the tempering experiments themselves.

A critique of Bartenev's theory in the light of experimental data over a wider range of operating variables was presented by Gardon (1965a), who applied Bartenev's approach for conditions other than "regular" by using a numerical method of calculating temperature histories and who also measured heat transfer coefficients independently of the tempering experiments. Temper stresses predicted by Bartenev's theory are exhibited in

Fig. 7 by the thin, dashed lines. This comparison with experimental results shows that, for high initial temperatures ($T_0 > 650°C$), Bartenev's theory also holds for moderately high quenching rates—up to about $h = 0.005$ cal/cm^2 °C sec, which produces "full temper" in 6-mm glass. However, for yet higher quenching coefficients, Bartenev's theory fits experiments progressively less well. It also does not fit temper stresses in glass quenched from relatively low initial temperatures or transient stresses while the glass is being cooled.

Meanwhile, Bartenev's theoretical work had also been criticized on more fundamental grounds. Thus Indenbom (1954) pointed out that the freezing-in of temperature gradients and the resulting definition of $\Phi(x)$ do not follow from the "instant freezing" hypothesis and are, in fact, inconsistent with the compatibility condition governing strains in the solidified portion of the glass. Still based on Bartenev's instant freezing hypothesis, Indenbom proceeded to a rigorous stress analysis of tempering.

Indenbom's point of departure was the recognition that the proper compatibility condition for a solidifying plate is not Eq. (7) but Eq. (2) and that—to handle the instant freezing assumption—Eq. (2) needs to be elaborated by considering separately the elastic, viscous, and thermal components of strain. Each of these is a function of x as well as t, even while the total strain [as per Eq. (2)] is a function of t only. The passage of time t can conveniently be represented by the progress of the freezing front X (Fig. 15), so that the compatibility condition for the solidifying plate may be written

$$\epsilon_{tot}(X) = \epsilon_{el}(x, X) + \epsilon_{vis}(x, X) + \beta T(x, X). \qquad (9a)$$

The "fluid" core of the plate that is above T_g will always conform to the deformation dictated by the frozen outer layers. Thus the viscous strain in any layer of the core will vary with time t (or with X) until the instant that layer freezes. Once a layer is frozen, the viscous strain in it cannot change any more: this frozen-in strain, also called the residual strain, is given by

$$\epsilon_{res}(x) = \epsilon_{vis}(X, X). \qquad (10)$$

For the frozen layers ($X < x < l$), the (variable) viscous strain of Eq. (9a) is replaced by the (constant) residual strain, so that

$$\epsilon_{tot}(X) = \epsilon_{el}(x, X) + \epsilon_{res}(x) + \beta T(x, X). \qquad (9b)$$

In Eq. (9b) the thermal strain, $\beta T(x, X)$, is presumed known, and there are three unknown strains. To solve the problem, two more equations are needed. One of these is provided by the instant freezing hypothesis, from which it follows that, at the instant of its freezing, any layer is stress-free. The elastic strain is therefore zero at $x = X$. Thus

$$\epsilon_{el}(X, X) = 0 \qquad (11a)$$

and, by definition,

$$T(X, X) = T_g. \tag{11b}$$

The other required equation follows from the equilibrium condition [Eq. (3)], which for the solid part of the plate can also be written

$$\overline{\epsilon_{el}(x, X)} = \frac{1}{l - X} \int_x^l \epsilon_{el}(x, X) \, dx = 0, \tag{12}$$

since stresses and elastic strains are proportional to one another.

Solving Eqs. (9b), (11), and (12), one obtains

$$\sigma(x, X) = \frac{E}{(1 - \nu)} [\overline{\epsilon_{res}(x)} - \epsilon_{res}(x) + \overline{\beta T(x, X)} - \beta T(x, X)]. \tag{13}$$

This is the solution for *transient* stresses in the glass, valid up to the instant of final solidification, i.e., for $X \leq l$ .When the glass has cooled to room temperature, average and local temperatures are the same, so that the *permanent,* or frozen-in stress distribution is given by

$$\sigma(x) = \frac{E}{(1 - \nu)} [\overline{\epsilon_{res}(x)} - \epsilon_{res}(x)]. \tag{14}$$

This is the general solution for permanent temper stresses, applicable for any symmetrical temperature–position–time history $T(x, t)$. For "regular" conditions of cooling, for which $T(x, t)$ may be approximated by the first term of a cosine series, Indenbom obtained the temper stress in the midplane of a plate as

$$\sigma_M = \sigma(0) = \frac{\beta E}{1 - \nu} (T_g - T_a) \left(1 - \frac{\sin \delta_1}{\delta_1} \right) \tag{15a}$$

$$\simeq \frac{\beta E}{1 - \nu} (T_g - T_a) \frac{\delta_1^2}{6} \left(1 - \frac{\delta_1^2}{20} \right), \tag{15b}$$

where δ_1 is the first root of

$$\delta \tan \delta = Bi = hl/k. \tag{16}$$

Equation (15b) is, to a first approximation, identical to the corresponding equation obtained by Bartenev. Indeed, for tempering with relatively low quenching coefficients, there is practically no difference between midplane stresses predicted by Indenbom's and Bartenev's equations. However, for higher quenching rates, i.e., for $\delta > \pi/4$, which for 6-mm glass corresponds to about 0.006 cal/cm² °C sec, Indenbom's theory is better, notably in that it predicts the attainment of plateau levels of temper at lower temperatures than does Bartenev's theory (cf. curves for $h > 0.006$ cal/cm² °C sec in Fig. 7). Kitaigorodskii and Indenbom (1956) also point to the superiority of

Indenbom's solution for predicting temper stresses in the surface (Fig. 9), which are, of course, the more important measure of the strengthening effect of tempering.* This distinction becomes especially important for the very high cooling rates encountered when glass is tempered in liquids (Boguslavskii *et al.*, 1964).

Here one might also note that, in view of its internal consistency, Indenbom's treatment, unlike Bartenev's, can be used to calculate transient stresses as well as permanent temper stresses. However, the transient stresses thus calculated bear no resemblance to the experimental data of Fig. 10. The reason for this is that the instant freezing hypothesis is too rough a description of stress relaxation for it to account for details of events in the transformation range. That, nevertheless, Bartenev's and Indenbom's theories yield as good results for many practical cases of tempering as they do is due mainly to the fact that temper stresses in glass quenched from relatively high temperatures are predominantly temperature equalization stresses, so that the precise rheological properties of glass in the transformation range are of relatively little moment.

Indenbom's theory is significant because it was the first rigorous treatment of tempering based on a physical model of solidification that, in principle, at least, allows one to comprehend solidification stresses as well as temperature equalization stresses. As such, it was the basis of Kalman's (1960) treatment of nonuniform tempering (Section V.A). In essence, Kalman (1959), Aggarwala and Saibel (1961), and Weymann (1962) also treated (uniform) tempering on the basis of Bartenev's instant freezing hypothesis.

D. "VISCOELASTIC" THEORY

The simplicity of the instant freezing theory lay in its characterization of glass as either an inviscid liquid (above T_g) or an elastic solid (below T_g). In fact, this transition occurs over an extended temperature interval, and glass in the transformation range exhibits both viscous and elastic characteristics. The first attempts at constructing a "viscoelastic" theory of tempering treated glass as a simple Maxwell body (Weymann, 1962; Indenbom and Vidro, 1964). However, the next important advance in tempering theory flowed from the development of computational methods of stress analysis in viscoelastic materials having more complex and, above all, temperature-dependent stress relaxation functions (Morland and Lee, 1960; Lee and Rogers, 1963). These methods were first applied to the tempering of glass by Lee *et al.* (1965), who treated glass as a thermorheologically simple, linear viscoelastic material and constructed a model of tempering based

---

*No closed form expression for surface stresses can be written even for the relatively restricted case of "regular" cooling.

on experimentally measured stress relaxation rates that had meanwhile also become available.

1. Characterization of Glass as a Thermorheologically Simple Viscoelastic Material

After Schwarzl and Staverman (1952), a viscoelastic material is said to be thermorheologically simple if the change with temperature of its relaxation modulus can be allowed for by a change of the time scale only. This can well be illustrated by the results of Kurkjian (1963), whose experiments on the relaxation of torsional stresses demonstrated that a *stabilized* soda-lime-silica glass, i.e., a glass brought to equilibrium at a given temperature, is indeed a thermorheologically simple viscoelastic material. That tempered glass, which is rapidly cooled, is not stabilized need not concern us for the moment.

Figure 16 shows Kurkjian's data: the decay with time of shear stresses

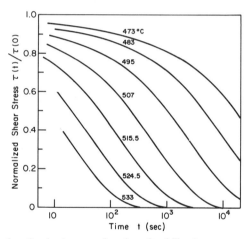

FIG. 16. Stress relaxation in glass as a function of stabilization temperature and time (Kurkjian, 1963).

induced in glass, stabilized at various temperatures, by a stepwise application of a shear strain at time $t = 0$. [The normalized stress is the actual stress at time t, divided by the instantaneous (elastic) stress produced at $t = 0$ upon straining the glass.] These curves illustrate three important aspects of the response of glass to being strained:

(i) The *instantaneous* response is taken to be elastic at all temperatures. This solid-like behavior lasts longer at lower temperatures. (At low enough temperatures, it continues long enough for glass to be regarded as an elastic solid.)

(ii) The *long-time* response of glass is to flow like a viscous liquid, allowing induced stresses to relax. Stress relaxation is perceptible the sooner, the hotter the glass. (At high enough temperatures, it starts soon enough for the glass to be treated as a viscous liquid.)

(iii) Finally, and specifically for stabilized glass, the stress relaxation curves, drawn on the basis of log(time), all have the same shape. This circumstance allows them to be brought into coincidence by shifting them along the log(time)-axis. The result is a "master" relaxation curve, such as that shown in Fig. 17, drawn for an arbitrarily selected base temperature of $T_B = 473°C$.

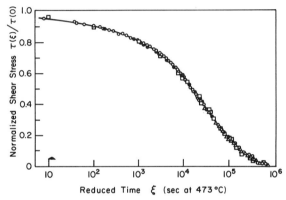

Fig. 17. Stress relaxation in stabilized glass as a function of "reduced time" (Kurkjian, 1963).

That the relaxation curves can be shifted in this manner shows that stabilized soda-lime glass is a thermorheologically simple viscoelastic material. The magnitude of the shift, by an amount $\Phi(T)$, is determined by the "shift function" ln Φ, given by

$$\ln \phi(T) = \ln \left(\frac{\eta}{\eta_B} \right) = \frac{H}{R_g} \left(\frac{1}{T} - \frac{1}{T_B} \right), \tag{17}$$

where η is the viscosity of the glass at temperature T [K], η_B the viscosity of the glass at an arbitrarily chosen base, or reference, temperature T_B [K], H the activation energy for viscous flow and stress relaxation (= 150 kcal/mol for soda-lime glasses), and R_g the universal gas constant (= 1.98 cal/mol K). Such a shift is equivalent to counting time faster at temperatures higher than T_B, and more slowly at temperatures lower than T_B. Following Lee *et al.* (1965), time as measured at the base temperature is called "reduced time" ξ. It is, of course, defined by

$$d\xi = \phi(T) \, dt. \tag{18}$$

2. The Stress–Strain Relation

The (time-dependent) shear stress $\tau(t)$ produced in a linear viscoelastic material by a stepwise applied shear strain of magnitude γ is

$$\tau(t) = G(t) \cdot \gamma, \tag{19}$$

where $G(t)$ is the time-dependent shear modulus, or "relaxation modulus," obtainable from Fig. 17 by multiplying its ordinates by $G(0)$, the instantaneous (elastic) shear modulus.*

Since Eq. (19) is linear in γ, the response of glass to a succession of strain increments may be obtained by Boltzmann's superposition principle. This yields a more general stress–strain relation in integral form

$$\tau(t) = \int_0^t G(t - t') \frac{d\gamma(t')}{dt'} \, dt', \tag{20a}$$

where t is the time of interest, t' a running time variable, from 0 to t, during which strain increments are accumulated at a rate $d\gamma(t')/dt'$, and $G(t - t')$, the "memory kernel," represents a measure of the "remembered" part of the corresponding stress increments.

If, in addition to varying strains, temperatures also vary, the reduced time ξ must be substituted for the real time t. This leads to the relation

$$\tau(\xi) = \int_0^\xi G(\xi - \xi') \frac{d\gamma(\xi')}{d\xi'} \, d\xi'. \tag{20b}$$

A similar relation may be written to relate tensile stresses $[\sigma(\xi)]$ to the tensile strain rate $[d\epsilon(\xi')/d\xi']$ and a time-dependent Young's modulus $[E(\xi)]$.

In contrast to the marked deformability of hot glass, its compressibility is very limited. The stress–strain relation for volume changes can therefore reasonably be characterized by a bulk modulus K that is independent of time and given, as for any elastic material, by

$$K = \frac{E(0)}{3(1 - 2\nu)} = \frac{2(1 + \nu)}{3(1 - 2\nu)} G(0).$$

The stresses produced in a plate by tempering are neither pure shear nor simple tension, but two-dimensionally isotropic tension [see Eq. (1)]. Muki and Sternberg (1961) derived an auxiliary relaxation modulus $R(\xi)$, which is for such a stress system what $G(\xi)$ and $E(\xi)$ are for pure shear and tension, respectively. $R(\xi)$ is defined by an integral equation in terms of K and either $G(\xi)$ or $E(\xi)$ (see also Lee and Rogers, 1963).

---

*$G(0)$ varies slightly with temperature in the transformation range, a variation neglected in the following treatment.

3. Calculation of Temper Stresses

Using the relaxation modulus $R(\xi)$, Lee *et al.* (1965) derived the following expression for transient stresses in tempering

$$\sigma(x,t) = 3 \int_0^t R(\xi(x, t) - \xi(x, t')) \frac{\partial}{\partial t'} [\epsilon(t') - \beta T(x, t')] \partial t'. \quad (21)$$

Note that in this equation the reduced time ξ is a function of position x as well as of time t, reflecting the x dependence of temperatures and therefore also of the shift function. Equation (21) incorporates both the stress–strain relation and compatibility condition [Eq. (2)]. From Eq. (21) and the equilibrium condition [Eq. (3)], the two unknowns $\sigma(x,t)$ and $\epsilon(t)$ can be obtained. For details of the derivation and numerical solution the reader is referred to Lee *et al.* (1965). In the following no more is attempted than to outline the procedure involved in this solution. For this it is convenient to define a new variable,

$$\sigma_g(x, t') = 3R(0)[\epsilon(t') - \beta T(x, t')]. \quad (22)$$

This represents the stress "generated" up to time t', regardless of any simultaneous relaxation. Consider next the *net* accumulation of stresses in a single layer of glass, for which Eq. (21) may be rewritten

$$\sigma(t) = \frac{1}{R(0)} \int_0^t R [\xi(t) - \xi(t')] \frac{\partial \sigma_g(t')}{\partial t'} dt'. \quad (23)$$

The step-by-step numerical solution of this for time $t = n \Delta t$ involves summing the remembered components of stress increments $\Delta\sigma_{gi}$ generated in n time intervals of duration Δt. Thus

$$\sigma_1 = \frac{1}{R(0)} [R_{11} \Delta\sigma_{g1}], \quad (24a)$$

$$\sigma_2 = \frac{1}{R(0)} [R_{21} \Delta\sigma_{g1} + R_{22} \Delta\sigma_{g2}], \quad (24b)$$

$$\sigma(t) = \sigma_n = \frac{1}{R(0)} [R_{n1} \Delta\sigma_{g1} + R_{n2} \Delta\sigma_{g2} + \cdots + R_{nn} \Delta\sigma_{gn}] \quad (24c)$$

The reason this calculation has to be performed step-by-step, rather than by proceeding directly to the nth time increment [Eq. (24c)], is that the solution for a single layer cannot be carried out independently of other layers. Rather, the total strain $\epsilon(t)$ must be determined at each intermediate time step, and stresses in all layers redistributed to satisfy the equilibrium condition at that instant. The individual $\Delta\sigma_{gi}$s thus represent the stresses generated by two simultaneous mechanisms:

(i) the direct generation of stress increments in response to small, step-wise changes in thermal strain, produced by the changing temperature distribution across the thickness of the glass; and

(ii) the redistribution of the sum total of remembered stresses in order that equilibrium be maintained across the thickness of the plate, even though stresses in different layers are relaxing at different rates.

These calculations necessarily require a high-speed computer. Even so, the step-by-step procedure outlined above is followed only as long as any part of the glass is still hot enough to relax. As, with decreasing temperatures, $R_{nn}/R(0)$ approaches its limiting value of unity to within the desired accuracy, these "viscoelastic" calculations are stopped. The stresses calculated to this point are what we have called solidification stresses. The further increment of stresses, from here to isothermal conditions at room temperature, are temperature equalization stresses in a wholly elastic plate, that can be calculated very simply by Eq. (4) of Section IV.B.

4. Results

Lee *et al.* did not pursue their analytical work to the point of comparing its predictions with experimental results. This was attempted by Narayanaswamy and Gardon (1969), who suggested a relatively minor modification of this treatment, bearing only on computational technique, in order to improve the precision with which Eq. (21) can be integrated. With that modification, they found the viscoelastic theory to be in good agreement with experimental data on temper stresses in glass quenched from high initial temperatures, but not low temperatures. This theory also yielded the first meaningful predictions of transient stresses during solidification. However, once again, agreement with experimental data was good for quenching from high initial temperatures only.

The "viscoelastic" theory thus appeared to be superior to the "instant freezing" theory only in that it could predict transient stresses as well as final temper stresses. In quenching from lower temperatures neither theory fitted the experimental facts. In 1969 Narayanaswamy and Gardon could but speculate that the reason for this was that, in quenching from low temperatures, some aspects of the glass transition that are not comprehended by the viscoelastic theory also become significant. In fact, this turned out not to be the only reason, as will be discussed in Section IV.F. Meanwhile, the next improvement in tempering theory was sought by taking account of the fact that stress relaxation during tempering is accompanied by structural changes in the glass, which also affect its density and viscosity; so that tempered glass, unlike stabilized glass, is not a thermorheologically simple viscoelastic material.

E. "STRUCTURAL" THEORY

The next, and for the present, last step in refining the theory of tempering is to take account of structural changes in the glass as it is being cooled through the transformation range. Such changes were thought likely to affect tempering in two ways: structure-related changes in viscosity would modulate the rates of stress relaxation already comprehended in the preceding "viscoelastic" theory, and structure-related changes in density would take a place alongside thermal strains as an additional driving force for tempering.

To allow for these effects requires a model of the relaxation, i.e., change with time, of the glass structure in a form that permits prediction of the evolution of structure-dependent properties for arbitrary temperature–time histories. Such a model was developed by Narayanaswamy (1971), using the mathematical formalism of Lee and Rogers (1963) for handling stress relaxation in viscoelastic materials with temperature-dependent properties.

1. Structure-Dependent Property Changes

When glass-forming liquids are cooled, their rapidly increasing viscosity inhibits their crystallization and causes them to retain the structure of a liquid, even while they become progressively more rigid. At low temperatures, i.e., in the "glassy" state, the structural arrangement of atoms is frozen, and thermal expansion or contraction reflect only changes in interatomic distances. These changes are instantaneous, reversible, and governed by the coefficient of expansion β_g for the glass. At much higher temperatures, the liquid is mobile enough for its volume changes also to be instantaneous and reversible. The corresponding, "liquid" coefficient of expansion β_l is larger than β_g for the reason that, in the liquid, temperature changes affect the *arrangement* of atoms as well as *distances* between them. At intermediate temperatures, i.e., in the transformation range, viscosities are so high that structural rearrangements cannot keep pace with temperature changes, but lag behind. The structure is then dependent on time (or cooling rate) as well as temperature; and the same goes for structure-related properties, such as density and viscosity.

Structural relaxation, i.e., the change with time of structure and structure-dependent properties, is inherently more complicated than stress relaxation: While the classic experiment on stress relaxation, namely, observation of the decay of stresses following a step change in strain, can be done isothermally and on stabilized glass (Section IV.D.1), the equivalent approach to studying structural relaxation necessarily entails a step function in temperature. Thus, even if one wishes to observe the relaxation only of, say, density, one cannot avoid simultaneous changes in other properties, including viscosity, which itself determines the rate of all relaxation processes, including that of density as well as stress. As a result, structural relaxation is inherently nonlinear, an effect first accounted for in this manner by Tool (1946).

Following Tool, it is helpful to characterize the structural state of the glass by its fictive temperature (see also Condon, 1954). This may be thought of as that temperature from which glass in equilibrium must be rapidly quenched in order to produce its particular structural state. If the glass is not quenched but cooled at some specified finite rate, its volume will change with temperature as shown schematically in Fig. 18. The con-

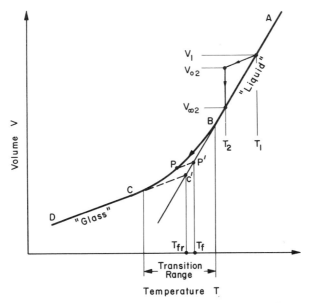

FIG. 18. Evolution of the specific volume of glass cooled at a given rate.

ventional definition of the (residual) fictive temperature T_{fr}, applicable below the transformation range, is given by the point of intersection (C') of the "glassy" and "liquid" lines. For the present purposes, the fictive temperature of the material needs to be defined not only in its final glassy state, but at all times as it passes through the transformation range, such as at the instant represented in Fig. 18 by the point P. Analogous to the residual fictive temperature T_{fr}, a time-dependent fictive temperature T_f is defined by the point P', where PP' is parallel to DCC'. PP' represents the glassy contraction of an equilibrium liquid at P' that is suddenly chilled to the instantaneous temperature of the material at P. It is evident from this definition that, above the transformation range, T_f is equal to the actual temperature T of the equilibrium liquid. T_f starts to lag behind T as the glass is cooled through the transformation range, reaching a limiting value of T_{fr} for all temperatures below the transformation range.

In the transformation range, physical properties such as specific volume and viscosity are thus functions not only of the actual temperature T, but also of the fictive temperature T_f, which, in turn, also depends on time. The glassy component of volume changes can reasonably be expected still to follow temperature changes instantly, while the structural rearrangement is delayed, tending to an equilibrium volume governed by the structural expansivity $\beta_s = \beta_l - \beta_g$.

Narayanaswamy (1971) accordingly defined a response function $M_v(\xi)$

$$M_v(\xi) = \frac{V - V_{\infty,2}}{V_{0,2} - V_{\infty,2}} = \frac{T_f - T_2}{T_1 - T_2}, \tag{25}$$

which gives the instantaneous specific volume V and corresponding instantaneous fictive temperature T_f as functions of the stepwise change in temperature $(T_1 - T_2)$ and the corresponding long-term change in the value of V from $V_{0,2}$ (immediately following the temperature change) to $V_{\infty,2}$ (its equilibrium value at the new temperature T_2). The reduced time ξ in Eq. (25) corresponds to the ξ used in Lee and Roger's (1963) treatment of nonisothermal stress relaxation. However, obtaining ξ from the real time t now entails a shift function that depends on T_f as well as T; so that Eqs. (17) and (18) are replaced by

$$\ln \phi(T, T_f) = \ln \left(\frac{\eta}{\eta_s} \right) = \frac{H}{2R_g} \left(\frac{1}{T} + \frac{1}{T_f} - \frac{2}{T_s} \right), \tag{26}$$

and

$$\xi = \int_0^t \phi(T, T_f) \, dt'. \tag{27}$$

For reasons to be discussed below, the strain point T_s is now taken as the base temperature in defining the shift function.

The coupling of Eqs. (25) and (27)—through the dependence of ξ on T_f, which is itself an unknown—is one consequence of the inherent nonlinearity of structural relaxation. Narayanaswamy's linearization of the problem by using a "fictive"-dependent shift function removes a major difficulty in analyzing phenomena of the glass transition, for it allows superposition. Thus, analogous to Eq. (20b), the general equations for the evolution of the specific volume and fictive temperature in response to an arbitrary temperature–time history $T(t)$ are

$$V(\xi) = V(0) \left[1 + 3\beta_l(T - T_0) - 3\beta_s \int_0^\xi M_v(\xi - \xi') \frac{dT}{d\xi'} \, d\xi' \right] \tag{28a}$$

and

$$T_f(\xi) = T(\xi) - \int_0^\xi M_v(\xi - \xi') \frac{dT}{d\xi'} d\xi'. \tag{28b}$$

The coupled Eqs. (26)–(28) remain to be solved for $V(t)$ and $T_f(t)$ by successive approximation.

The response function $M_v(\xi)$ may be obtained from volume relaxation experiments in a manner analogous to obtaining stress relaxation moduli. The results of Rekhson and Mazurin (1974) can be represented by the empirical expression

$$M_v(\xi) = \exp(-\xi/\tau_v)^{0.68}, \tag{29}$$

where τ_v is the "characteristic" volume relaxation time (in which M_v falls to $1/e$) and which is 6345 sec for soda-lime glass at its strain point.

Here it might be noted that, since Eq. (29) is not a simple exponential, it, unlike Tool's (1946) equation, comprehends the effects of "memory" as well as the nonlinearity of relaxation (see Goldstein, 1964). However, in tempering it is mainly the nonlinearity, due to variations in viscosity, that is important; and, as will be seen in Section IV.E.3, the area under the relaxation curve is more critical than its exact shape. Thus Tool's equation would have sufficed to allow for structural effects in tempering, and a model based on this was in due course also constructed (Ohlberg and Woo, 1974). Memory, i.e., the effect of prior thermal history on the direction and rate of change of structural state, is more likely to be important in annealing (Gardon and Narayanaswamy, 1970) and in other heat treatments, such as "temnization" (Acloque and Pèyches, 1954), in which temperature is not a monotonically decreasing function of time (cf. the "crossover" experiments of Macedo and Napolitano, 1967).

2. Calculation of Structural Effects in Tempering

Narayanaswamy's scheme for computing stresses by the "structural" model of tempering is basically the same as that for the viscoelastic model, outlined in Section IV.D.3, only now the evolution of a structure-dependent viscosity and density must be followed as well as that of stresses. Structural changes in density are taken into account along with ordinary thermal expansion by rewriting Eq. (21) of the viscoelastic theory with an additional term:

$$\sigma(x, t) = \int_0^t R\left(\xi(x, t) - \xi(x, t')\right)$$
$$\times \frac{\partial}{\partial t'}\left[\epsilon(t') - \beta_g T(x, t') - \beta_s T_f(x, t')\right] dt'. \tag{30}$$

The effect of structural changes in viscosity on the stress–time history is allowed for by expanding or contracting the time scale of the modulus R not only in response to temperature changes but also in response to changes in the fictive temperature T_f [as per Eq. (27)]. Given the temperature history $T(x, t)$ of the glass, the required history of fictive temperatures $T_f(x, t)$ must therefore be calculated first. This requires an iterative solution, by successive approximations, of the coupled Eqs. (26), (27), and (28b). Once both $T(x, t)$ and $T_f(x, t)$ are known for all times, Eq. (30) can be solved for $\sigma(x, t)$, following the step-by-step procedure schematically indicated for the viscoelastic case by Eqs. (24).

3. Results

At first, predictions of the structural theory agreed well with experimental tempering data only for quenching from relatively high initial temperatures. As was the case with the viscoelastic theory, agreement was rather poor with data obtained for low T_0s, for which the degree of temper is a strong function of T_0. The key to resolving this difficulty was Narayanaswamy's (1978) recognition that the viscosity–temperature relation implied by the stress relaxation moduli used in the calculations may not be exact for the glass being tempered.* Under the circumstances, the exact viscosity–temperature relation—which is given by areas under the $G(t)$ curves for various temperatures—is more important for accurately predicting stresses produced by quenching from low temperatures than is the precise shape of those curves. Relaxation moduli having been measured for only a few glasses; any difference between their viscosity–temperature relation and that of the glass being tempered must therefore be allowed for. This can readily be done by using strain points as the base temperatures for defining the reduced time ξ and adjusting all shift functions accordingly.†

With this adjustment made to independently obtained stress and volume relaxation data (Kurkjian, 1963; Rekhson and Mazurin, 1974), Narayanaswamy showed that the structural theory of tempering comes into good agreement with all relevant experimental data: those on transient stresses as well as permanent stresses, and for low initial temperatures as well as high. It is thus superior to all preceding theories, and it is, of course, the

---

*At any given temperature, the viscosity η and shear relaxation modulus $G(t)$ of a stabilized glass are related by

$$\eta = \int_0^\infty G(t)\, dt.$$

†The strain point T_s is the temperature at which the viscosity of a stabilized glass is $10^{14.5}$ poise. Since the activation energies of similar glasses are substantially identical, T_s serves well to fix the viscosity scale of a given glass relative to the temperature scale. Retrospectively, it is clear that this procedure is also applicable for the viscoelastic theory.

only one that can comprehend the development of a nonuniform density distribution in tempered glass (Fig. 11).

Even so, a direct comparison of theoretical predictions and experimental observations does not show unequivocally how structural changes affect temper stresses. However, this question could be explored more definitively, even if indirectly, by some "computer experiments" on hypothetical materials in which viscosity or density changes, which in nature occur together, are deliberately and individually suppressed in order to learn how various manifestations of the glass transition affect the tempering process.

The results are shown in Fig. 19. This shows that structural changes do

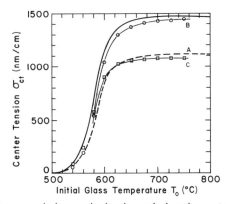

FIG. 19. Effect of structural changes in density and viscosity on temper stresses.————, structural model; –––, viscoelastic model; ⊙⊙⊙, structure dependence of viscosity suppressed; ⊟⊟⊟, structure dependence of density suppressed. Glass thickness, $L = 0.61$ cm; heat transfer coefficient, $h = 0.0045$ cal/cm^2 °C sec (Narayanaswamy, 1978).

indeed have a marked effect on the degree of temper. Primarily, this effect is one of structural volume changes. Thus the suppression of structural viscosity changes (curve B) yields substantially the same results as full allowance for structural changes. Conversely, calculations in which structural effects on viscosity are retained, while density changes are suppressed, yield essentially the same σ_M versus T_0 curve (C) as the viscoelastic theory (curve A). For the effect of these structural density changes upon temper stresses to be comprehended by the viscoelastic theory, it would have to be used with an "effective" expansivity of β' higher than β_g, as will be discussed below.

For the case illustrated by Fig. 19, which for $T_0 > 620$°C corresponds to full temper in 6-mm glass, structural effects account for 24% of the permanent temper stress produced. The structural contribution to temper stresses varies inversely with the severity of quenching: it is greatest for relatively slowly cooled (or annealed) glass, where it amounts to 40% of the total, and least for the most severely quenched glass.

This last conclusion is contrary to that of Indenbom and Vidro (1964) (cf. also Boguslavskii and Pukhlik, 1969), who, from their treatment of "thermoplastic" (i.e., viscoelastic, using the Maxwell model) and structural stresses in tempered glass, concluded that the relative importance of the latter increases with increasing severity of quenching. However, Indenbom and Vidro calculated what they called structural stresses simply from the terminal distribution of densities in the glass, without allowing for the fact that stresses of structural origin, like those of thermal origin, develop in a temperature interval in which stress relaxation also occurs. Residual structural stresses do not, therefore, necessarily reflect the full range of the terminal density distribution in tempered glass.

Conversely, it is interesting to note here that stresses of structural origin have also been identified in annealed glass, which is, of course, structurally homogeneous at the end of its heat treatment (Gardon and Narayanaswamy, 1970). This clearly shows that permanent stresses of structural origin can be caused—and are, in part at least, caused—by *transient* non-uniformity of structure (or fictive temperature), regardless of the terminal homogeneity or heterogeneity of the structure of the heat-treated glass.

F. Assessment of Theories

Perhaps the first point to be made in comparing the three theories of tempering just reviewed is that they are closely related. Thus Indenbom's treatment of the "instant freezing" hypothesis can be recognized as a special case of the "viscoelastic" theory, which, in turn, is a special case of the "structural" theory. Starting with the last of these, the viscoelastic theory may be derived by suppressing all structure-related changes in physical properties, i.e., by setting $T_f \equiv T$ or $\beta_s \equiv 0$. Indenbom's theory [see Eq. (13)] can be derived from that of Lee *et al.* by setting $R(\xi(x, t) - \xi(x, t'))$ in Eq. (21) identically equal to zero for $t' < t_g$, and equal to $R(0) = E(0)/3(1 - \nu)$ for $t' > t_g$, where t_g is the time at which $T(x, t_g) = T_g$.

Discrepancies between experimental observations and the predictions of earlier theories of tempering could be ascribed to incompleteness of the theories. Thus it was expected that shortcomings of the instant freezing theory would be corrected by the viscoelastic theory, and its shortcomings, in turn, by allowance for structural effects. To some extent these expectations have been fulfilled; but, in some respects, predictions of the later theories are no better than those of the first. One reason for this is that tests of these theories also constitute tests of the available data on relevant physical properties of glass. Shortcomings of these data can be more clearly recognized now, with the availability of a comprehensive theory and attempts to interpret tempering data using physical properties obtained from independent experiments.

Here it should be noted that, even if all relevant physical properties were known, they could be used only with the structural theory, since this is the only one that refers to a "real" glass. The other theories necessarily require some "effective" physical properties. In light of the structural theory one can recognize, for example, that the coefficient of expansion β' to be used with the viscoelastic theory should lie between β_g and β_l, the latter being typically three times as large as β_g (Haggerty and Cooper, 1965). Thus, depending on what aspect of tempering one is considering and what data one has available, it is possible for the simpler theory to yield the better predictions. Indeed, the instant freezing theory is not only the simplest to use, but for restricted operating parameters, at least, it can also predict final temper stresses as well as the more sophisticated theories. All can, of course, be improved by a judicious choice of "physical properties," notably by deriving these from the tempering experiments themselves.

Thus, if one treats β' as an adjustable parameter, the viscoelastic theory can be made to fit experimental data on plateau level temper stresses perfectly. (Narayanaswamy, 1978). The effective values of β' range from $1.5\beta_g$ for relatively slow cooling (by natural convection) to only $1.2\beta_g$ for severe quenching. These figures reflect structural contributions to permanent temper stresses of from 34 to 17%, respectively. If, in addition—as was found desirable in the case of the structural theory (see Section IV.E.3)—the time scale of the relaxation modulus is shifted so as to adjust for the strain point of the glass in question, predictions of the viscoelastic theory are greatly improved for low initial temperatures also. Indeed, used in this manner, the viscoelastic theory can be made to fit experimental data on permanent temper stresses perfectly: better even than the structural theory using independently measured physical properties.

In summary:

1. Structural contributions to permanent temper stresses depend on the cooling rate. They are most marked for relatively low cooling rates, where they may account for 30–40% of the total residual stress; and they become progressively less significant for more intensely quenched glass.

Structural effects on temper stresses are well comprehended by the "structural" theory, which is also the only theory geared to using property data on "real" glasses. To allow for structural effects when using the "instant freezing" or "viscoelastic" theories requires the use of "adjusted" expansion coefficients, as has always been the case in classical annealing theory (Gardon and Narayanaswamy, 1970).

2. Subject to the above limitation in the case of the instant freezing and

viscoelastic theories, all theories do reasonably well in predicting the plateau levels of temper stresses obtained when glass is quenched from high initial temperatures at low rates. This is not a searching test, however, since the temper stresses involved are almost entirely temperature equalization stresses, and Adams and Williamson's (1920) theory of annealing would predict them about as well.

Still in the plateau regime, i.e., for high T_0s, but for higher heat transfer coefficients, solidification stresses become progressively more important, and the viscoelastic and structural models become preferable to the instant freezing model.

3. Accuracy in predicting plateau levels of degree of temper is limited to about 20% by uncertainty regarding heat transfer coefficients and some properties of the glass, notably its heat capacity and thermal conductivity in the transformation range and the expansivities β_g and β_l, not to mention β'. Predictive capability for surface stresses is lower than that for midplane stresses, partly on account of greater uncertainty in calculating surface temperatures.

4. For glass quenched from low initial temperatures, i.e., in the regime in which temper stresses are strongly dependent on T_0, the instant freezing model ceases to be meaningful even before $T_0 \leqslant T_g$. In using either of the other two theories, the critical factor appears to be the accord of the stress relaxation data used with the strain point of the glass.

5. Transient stresses in the transformation range, i.e., during solidification of the glass, are quite small, usually only a few percent of the final temper stresses. They are of some interest as indicators of the propensity of the glass to break during processing and as means to test the predictive capability of various models of the heat transfer and stress relaxation processes involved.

Transient stresses cannot be calculated at all by Bartenev's theory. They can be calculated by Indenbom's theory, but the results are misleading, for the instant freezing assumption allows for too much stress relaxation early in cooling, and none at all later. This leads to an underestimate of transient surface tensions, which, if they occur at all, do so early during quenching (Fig. 10). Its ability to predict transient stresses is perhaps the most searching test of the structural theory of tempering. It does remarkably well even for low T_0s, for which the viscoelastic theory fails (Narayanaswamy, 1978).

6. The heterogeneity of (stress-free) densities in tempered glass, which is the most direct manifestation of structural changes in the glass during tempering, can, of course, be comprehended by the structural theory only. Its ability to predict these is illustrated in Fig. 11 by the agreement of calculated density distributions (point symbols) with experimental data (continuous plots).

V. Nonuniform Tempering

A variety of situations can arise in which, incidentally or deliberately, the tempering of glass is conducted in a nonuniform manner. Thus, unequal quenching of opposite surfaces of the glass may be employed to deform initially flat glass into a curved shell. Lateral nonuniformity of heat treatment may produce temper stresses that vary over the extent of a plate and also give rise to what we shall call membrane stresses.

A. MEMBRANE STRESSES IN TEMPERED GLASS

The temper stresses that have been considered so far are two-dimensionally isotropic stresses in the plane of a plate, uniform over its lateral extent and varying through its thickness only. We shall now call them "ordinary" temper stresses to distinguish them from "membrane" stresses, which may also be produced by tempering, and which, by definition, are also stresses in the plane of the plate but constant through its thickness. Just as ordinary temper stresses result from transient temperature distributions that vary though the thickness of a glass plate, so permanent membrane stresses may be produced by transient temperature distributions along its lateral and/or longitudinal dimensions. In general, ordinary temper stresses and membrane temper stresses will coexist. Membrane stresses will be zero in central regions of a large plate uniformly quenched from a uniform initial temperature. However, even in such a plate the almost unavoidable preferential cooling of edge regions is likely to produce compressive stresses parallel to the edges. These edge stresses are, in effect, one-dimensional membrane stresses and the most widespread manifestation of permanent membrane stresses in glass, tempered or annealed (see Section V.B).

Edge stresses are usually compressive and therefore welcome even though produced only incidentally. Other incidentally produced membrane stresses may be less welcome. Thus, spatial nonuniformity of cooling, encountered to some extent with most quenching systems, often leads to membrane stresses the variation of which reflects the spacing of the cooling jets. These relatively small stress fluctuations may have little or no effect on the strength of the glass. They are nonetheless unwelcome for, under certain lighting conditions, they impart to the glass an objectionable iridescence or mottled appearance (see Fig. 27 and p. 208).

Apart from these two examples of membrane stresses that arise incidentally, membrane stresses are also being produced deliberately to achieve certain special effects. Some examples are considered in Section V.C.

The physical mechanisms that govern the generation of ordinary temper stresses (see Section III) also govern the generation of permanent membrane stresses. Both have their origins in transient thermal strains, the (partial) relaxation of which leads to permanent stresses in the glass. Both are systems of internally balanced plane stresses; only that—while ordinary temper stresses have to balance across the thickness of the glass—membrane stresses, which may vary in two dimensions, have to balance across any longitudinal or transverse section of the plate. The two stress systems also differ as to the extent to which their distributions may be controlled. We have seen that the distribution of ordinary temper stresses through the thickness of the glass is always substantially parabolic, since it is uniquely determined by heat fluxes within the glass that can be controlled only at the surfaces. Thus, while the magnitude of temper stresses can readily be controlled, one has almost no control over their spatial distribution. In contrast, both the magnitude *and the shape* of the lateral and longitudinal distributions of permanent membrane stresses may, within limits, be arbitrarily chosen, since the transient temperatures that produce them can, in principle, be controlled over the entire extent of the plate by a suitable choice of T_0 and h as functions of position (y, z).

Several factors complicate the calculation of permanent membrane stresses. First of all, ordinary temper stresses vary in one dimension only, and they could usefully be analyzed without reference to edge effects or even to any slight curvature of the plate to be tempered. In contrast, membrane stresses are intrinsically two dimensional, for even where they are caused by nonuniform temperature distributions in one dimension only, the maintenance of equilibrium will generally require stresses in the second dimension also, along the edges of a plate, at least. Thus, unlike ordinary temper stresses, membrane stresses must be considered in the context of the size and shape of the entire plate. This, together with the great variety of areal temperature distributions that may be involved, makes the solution of any membrane stress problem inherently less general than the corresponding treatment for ordinary temper stresses. The calculation of membrane stresses is further complicated by the fact that membrane stresses, unlike ordinary temper stresses, may affect the curvature of a plate. Under the circumstances, it is not surprising that, for all the inventions in this area, only one theoretical treatment of permanent membrane stresses has appeared in the literature (Kalman, 1960).

Kalman was interested in producing a tempered television faceplate, the edges of which were to be left untempered in order that they might be heat-sealed to other parts of a novel TV tube. He recognized that such nonuniform tempering would produce membrane stresses as well, and he calculated their magnitude on the basis of Bartenev's instant freezing model of the solidification of glass. Kalman's calculations were, in some ways, analogous to Indenbom's: where Indenbom, in calculating ordinary temper stresses, followed the progress of the freezing front through the thickness of a glass plate, Kalman's calculation of membrane stresses followed the progress of the freezing front in the other two dimensions of the plate. In fact, the freezing front in a three-dimensional object will generally be a three-dimensional surface (X, Y, Z). By considering its position in Y and Z only, one in effect calculates membrane stresses that are superimposed on ordinary temper stresses. The total stress at any position within the glass is then given by the local sum of the two stress systems, as indicated in Fig. 20. Where the membrane stress is tensile, it detracts from the com-

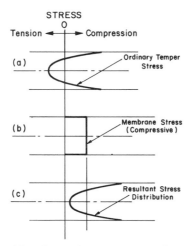

FIG. 20. Superposition of a membrane stress on ordinary temper stresses.

pressive ordinary temper stress in the surface, thus weakening the glass. Where, as in Fig. 20, the membrane stress is compressive and larger than the tensile ordinary temper stress in the midplane, it suppresses the propensity for spontaneous crack propagation in tempered glass. (see Sections I.A and V.C.1).

While Kalman was interested in membrane stresses that accompanied the production of spatially varying ordinary temper stresses, membrane stresses may also be superimposed on ordinary temper stresses that are spatially uniform. The difference between such stress systems is illustrated by two examples:

Case 1. h uniform, T_0 varied. Consider a glass plate divided into two regions, A and B. Both regions are quenched with the same heat transfer coefficient h, but from different initial temperatures T_0. If both T_{0A} and T_{0B} are high enough, the ordinary temper stresses produced in the two regions will be equal to one another and to the "plateau-level" of temper corresponding to the given heat transfer coefficient (see Fig. 7). The only effect of different initial temperatures will be that the region quenched from the lower T_0 will solidify ahead of the other. Analogous to the simplest view of how ordinary temper stresses are generated (Section III.A), the region that solidified first will tend to go into compression, while the other goes into tension. The actual stresses generated in regions A and B will, of course, depend on their configuration and also upon their positions relative to the edges of the glass. If A is a circular region at the center of a larger circular plate, and if $T_{0A} < T_{0B}$, such tempering will put region A under uniform, isotropic compression, balanced by a hoop stress in the annular region B. The resulting distribution of radial and tangential membrane stresses is indicated schematically in Fig. 21, along with the uniform ordinary temper produced.

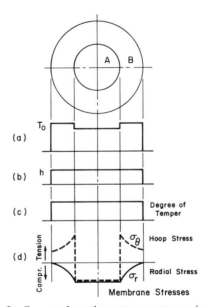

FIG. 21. Generation of ordinary and membrane temper stresses in a circular disk, case 1.

Case 2. T_0 uniform, h varied. If, instead of the previous arrangement, one starts from a uniform initial temperature but with $h_A > h_B$, region A will again solidify before region B. This will produce membrane stresses similar to those considered in case 1. Now, however, the use of different quenching coefficients also results in a difference between the degrees of temper produced in the two regions, and—unlike the situation illustrated in Fig. 21c—the ordinary temper in region A will be higher than in the surrounding region B.

B. EDGE STRESSES

Stresses in the edges of glass plates, tempered or annealed, are the most widespread manifestation of permanent membrane stresses in glass. They are usually compressive, and therefore generally beneficial, especially in annealed glass. They are produced incidentally, without any effort by the operator, simply as a result of a locally higher surface-to-volume ratio that usually causes the edges of a plate to cool more rapidly than its interior. The edges of the glass thus solidify before the interior, which can accommodate itself to the contraction of the edges. When, later on, the interior also solidifies and contracts, this is resisted by the already solid edge regions, which are thus put into compression. Compressive forces in a narrow edge zone are balanced by tensile forces in the interior (see Fig. 27). In curved plates the balancing tensile forces tend also to be confined to a relatively narrow band near the edge.

Edge compression may be enhanced by more intensive quenching of the edges (Ermlich, 1965). Similarly, tensile stresses may also be produced in the edges by deliberately shielding them from quenching. While deliberate action may thus produce either tensile or compressive stresses in the edges, the complete avoidance of membrane stresses in the edges is virtually impossible. The reason for this is that, even in the absence of any local temperature gradients that *directly* produce permanent stresses in a given edge, stresses are likely to arise in it in response to, and to equilibrate, thermal strains imposed in other parts of the plate. This effect is well illustrated by membrane stresses in a ribbon of float glass that has been annealed in the presence of a laterally nonuniform temperature distribution. Far from its end, membrane stresses in such a continuous ribbon are purely longitudinal. No transverse stresses are produced, since longitudinal temperature gradients are virtually negligible. Yet, the end of the ribbon, and the ends of large plates cut from it, become the seats of membrane stresses parallel to these edges, i.e., transverse to ribbon.

C. APPLICATIONS OF NONUNIFORM TEMPERING

1. Nonuniformly Tempered Windshields

Deliberate nonuniform tempering is practiced primarily in the manufacture of tempered windshields that have fracture characteristics favorable for the maintenance of vision after breakage. Since no tempered windshields are allowed in the U.S. and Canada, these are all European developments, discussed here primarily to illustrate some remarkable effects that can be obtained by nonuniform tempering.

One problem with conventionally tempered windshields is that, if broken *and* retained in their frames, they tend to obstruct vision by the fineness of their "fracture pattern" (see Fig. 2), which is the very feature desired to make fragments safe in case the glass is *not* retained in its frame. The obstruction of vision is especially marked with windshields installed obliquely to the line of sight and for lighting conditions under which light scattered from the many fracture surfaces veils the exterior scene. The inventors of a variety of nonuniformly tempered windshields have sought to maintain good visibility through broken glass by providing for selected regions either to remain unbroken or to break into coarser particles. The fracture patterns of some of these windshields are illustrated schematically in Fig. 22.

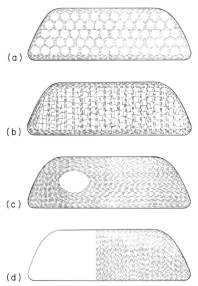

FIG. 22. Schematic representation of the fracture patterns of various differentially tempered windshields. (a) Sekurit W; (b) Sekurit BS or Zebra-zone glass; (c) Visurit; (d) Bisekurit.

In two forms of nonuniformly tempered glass, some regions are less strongly tempered than others, in order that—when the glass breaks—the former regions shall not dice. In the earliest form of this glass, "Sekurit W" (Long, 1941), many circular regions, 4–5 cm in diameter, were left essentially untempered. When broken, this glass had the appearance sketched in Fig. 22a. The unbroken domains continued to afford the driver vision even if the glass was retained in the vehicle. If it was not, the size of fragments presented some hazard to occupants, and for this reason Sekurit W is no longer used.

In a newer form of this type of windshield, known as "zebra glass" in Britain and as "Sekurit BS" in Germany, the "reserved zones" are mildly tempered, so that they still dice when broken, but into relatively coarse particles (St. Gobain, 1966). Thus, the size of fragments to which occupants might be exposed is less than in the previous case. At the same time, the arrangement of these coarser fragments in vertical rows still permits some vision even through obliquely installed windshields (Fig. 22b). Nonuniform temper of this type is produced by quenching with spatially varying heat transfer coefficients (Jochim, 1965; Baker, 1967; Bunn, 1967).

In two other types of windshield, the glass is fully tempered over its entire extent, and vision after breakage is assured by restricting the propagation of fractures to one side or the other of an invisible barrier produced by differential tempering. The barrier is a region in which compressive membrane stresses are high enough to more than offset the midplane tension of ordinary tempering, putting the glass under compression throughout its thickness (Fig. 20). Since the spontaneous propagation of fracture in tempered glass depends on the presence of tensile stresses in its interior (see Section I.A), fractures will not propagate across such a barrier. Dicing is thus restricted to the side in which fracture was initiated.

In the case of "Visurit" the barrier separates a circular "window" in front of the driver from the rest of the windshield. Both zones are fully tempered and an annular boundary region between the two, about 2 cm wide, is under tangential compression (Chan and Lambert, 1963). Should a stone hit the window, vision is assured through the rest of the windshield. If, as is more likely, the larger part of the windshield is broken, vision is assured through the window, as shown in Fig. 22c. The drawback with Visurit is that the unbroken window is large enough to constitute a hazard to the driver, should it be forced into the vehicle by pressure on the broken windshield.

This danger is avoided in yet another form of differentially tempered windshield, "Bisekurit," which is divided into two zones as shown in Fig. 22d (Long, 1958). Now, if either of the two zones is broken, continued vision is assured through the other and, at the same time, the unbroken part of the windshield is always retained in its frame. Bisekurit would thus seem to be the answer to both problems: that of vision if the glass is retained in its frame, and that of safety if it is not retained. However, it has not become established commercially. One technical problem is that it is virtually impossible to maintain compressive stresses along the entire length of a barrier that, unlike the one in Visurit, is not a closed loop, but extends from one edge of the windshield to the other. The probability therefore exists that, even though with some delay, fractures will propagate around the ends of the barrier, which also are potentially vulnerable parts of the windshield.

Differentially tempered glass of the type of Visurit or Bisekurit can be made by varying T_0 or h, or both, over the extent of the plates. Ways of doing this are described by Chan and Lambert (1959), Bertrand and Acloque (1964), Long (1965b), and Acloque (1966).

One drawback with all these nonuniformly tempered windshields is that the marked variation of membrane stresses impresses on them marked photoelastic patterns, as shown in Fig. 28, which was obtained in polarized light. In partially polarized natural light, such as skylight, these are less pronounced, but still quite noticeable even by the unaided eye. To a wearer of polarized sunglasses the photoelastic stress patterns may, under certain circumstances, become so bright as to veil outward vision.

2. Unsymmetric Heat Treatments and Thermally Warped Glass

An entirely different form of thermal tempering retains uniform quenching over the extent of a plate but introduces a difference in the intensity of quenching applied to the two sides. This would tend to freeze-in an asymmetric distribution of stresses through the thickness of the plate. This is inherently unstable. It will resolve itself into two components, one of which is a *linear* stress gradient through the thickness of the glass, which can relieve itself by bowing an initially flat plate into an approximately spherical shell (Carson *et al.*, 1968; Dennison and Rigby, 1970). With this component of temper stresses relieved by flexure, the temper stresses remaining in the now curved plate will be balanced with respect to both forces and bending moments across any section of the plate. Figure 23 illustrates the resolution of a potentially unsymmetric distribution of temper stresses (curve a) into bending stresses (b), relieved by deformation of the plate, and the residual temper stresses (c).

In some horizontal tempering processes, in which one side of the glass may run a little colder than the other, differential quenching of the two sides is also used to offset this temperature difference and thus maintain flatness of the glass (Shabanov *et al.*, 1970).

VI. Tempering Practice

A. HISTORICAL OVERVIEW[*]

The first manifestation of anything like "tempered glass" may well have been what since the 17th century have been called Prince Rupert's drops. These were scientific curiosities, obtained by letting hot, very fluid glass drip into water. The result is a tear-shaped drop with a long fiber for a tail.

---

[*]For bibliographies on the early history of tempered glass, see Duncan (1960) and Schultze (1931).

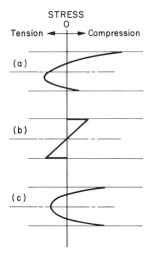

FIG. 23. Stresses in thermally warped tempered glass. (a) With the upper surface quenched more severely, an asymmetric stress distribution would be produced in a plate constrained to remain flat. (b) Since the plate is not so constrained, a *linear* stress distribution can be relieved by bending. (c) Represents the residual stress distribution in the bent plate. Its upper surface is now convex.

The quenching of glass from so high a temperature into water is about as intense as can be obtained, and, as a result, the body of a Prince Rupert drop is extremely strong. It cannot usually be broken even with a hammer. The tail, on the other hand, is vulnerable and can readily be broken off. This releases the strain energy in this very highly tempered piece of glass, which is reduced into very fine, powdery fragments.

In the 1870s several inventors attempted to put this phenomenon to use by strengthening glass objects having more useful shapes. This required quenching the glass from much lower temperatures in order to maintain form, and at much lower rates to prevent breakage in the process. Two principal approaches were tried: quenching in liquids (de la Bastie, 1875) and quenching by contact with solid platens (Siemens, 1877)*. (See also Henrivaux, 1883). At the time, the use of tempered glass was very actively promoted for a wide variety of products, including even railroad ties—a promotion of which the beginning seems to have been better documented than the end. For more established glass products, like goblets, de la Bastie's liquid quench had the virtue of allowing the inner as well as outer

---

*For ease of reference, U.S. Patents have been cited in preference to foreign patents, at some risk to priority of dates. In other instances, patents have been cited for their description of the state-of-the-art rather than their particular contribution.

surfaces to hollow objects to be tempered. Its great drawback was the need to use liquids that had only moderate heat transfer capability, such as mixtures of oils and greases, and even these had to be preheated in order to reduce yet further the rate at which heat was extracted from the glass. Quenching red-hot glass into preheated oil must have been a rather disagreeable and dangerous operation.

Siemens aimed more specifically at producing tempered flat glass, which could not be made by liquid quenching without suffering some distortion. He combined the forming and tempering of glass disks in one operation by pressing glass between cold platens of clay or metal. This process produced glass plates of good form but with marred surfaces. For applications requiring transparency, Siemens's glass had to be ground and polished after tempering, and, in turn, the removal of surface layers diminished the remaining surface compression. However, grinding and polishing individual plates was less of a drawback a hundred years ago than it would be today, with the availability of good quality flat glass made by simpler processes, and Siemens's technique appears to have been used extensively for making sight glasses for boilers and windows for portholes.

Tempering did not become a process of great industrial potential until ways were found to preserve not only the shape but also the surface finish of preformed glass articles. This was accomplished for flat glass with the invention in France in the 1920s of ways to quench glass at adequately high rates with impinging air jets, which no longer marred the glass surface. This technology and some of its contemporary developments are discussed in greater detail in the following section. Thermally tempered safety glass started to replace laminated safety glass in European automoblies in the early 1930s. In American automobiles tempered safety glass has been used for all windows, except windshields, since the mid-1960s. More recently, Federal regulations have also made tempered safety glass mandatory in many architectural applications. Tempered flat glass, including some mildly curved plates such as are used in automotive glazing, now constitute the largest part of the tempered glass market, amounting to approximately 500 million dollars per year in the U.S. alone.

Other widespread applications of thermally tempered glass include safety eyeglasses, sight glasses for pressure gauges, and glass piping and reaction vessels for some chemical processes. Mainly because of the difficulty of adequately quenching the interior of hollow ware, tempering has not been applied to bottles and drinking glasses. On the other hand, the rims of some tumblers are now heat strengthened, which makes them somewhat more resistant to chipping without imparting to them the strength and fracture characteristics of fully tempered glass.

Interestingly enough, after the great advances made in tempering by air quenching, increasing attention is again being given to quenching in liquids and by contact with solid platens. New approaches to these older technologies are briefly touched upon in Sections VI.B.4 and VI.C.

B. FLAT GLASS

1. Vertical Air Quenching

The widespread use of tempered glass began with the invention of a practical system for quenching glass with impinging jets of air. Many elements of this invention are still practiced, although there have also been some radical improvements.

In the vertical tempering operation, flat glass is suspended near its upper edges by tongs, resembling miniature ice tongs. Thus, once gripped, the glass is held the more tightly the heavier the pull. To ensure a good grip even on cold and hard glass, the tips of the tongs are pointed and hardened. To avoid excessive deformation of the hot glass in the vicinity of the tongs, their number is chosen with due regard to the weight of the part carried. Whenever the number is greater than two, precautions need to be taken to ensure a uniform distribution of the load and to avoid any tendency to bend the glass.

The glass thus suspended is sent through a vertical furnace. To ensure uniform heating over the extent of the glass, heating is conducted primarily by radiation. Furnaces may be electrically heated, with heating elements well distributed over the vertical walls. Gas furnaces may employ extended surface combustion, obtained by admitting the gas through porous bricks, or arrays of high intensity "radiant cups."

To avoid deformation of the glass, its residence time at higher temperatures must be limited. This precludes letting the glass come to equilibrium in a furnace maintained at the temperature desired for subsequent quenching. Instead, furnaces—or the final zones of multizone furnaces—are usually operated from 50 to 100°C above the desired exit temperature of the glass, which must then be removed from the furnace at just the right time. Furnaces for mass production are run on timed cycles, and the size and thickness of the parts fed to them must therefore be controlled. Alternatively, the temperature of the glass in the furnace may be monitored in order that it may be withdrawn when it has reached the desired level.

Upon reaching the desired temperature, the glass is positioned between two vertical "quench heads" and rapidly chilled by air jets that impinge on it from both sides. The first quench heads (St. Gobain, 1931) consisted of grids of iron pipes with small (1–2 mm) holes drilled into their sides. They were supplied with compressed air at 4–6 atm pressure, and the "spent" air escaped between the grids. The local heat transfer rates produced by such small, high-velocity air jets are highly nonuniform (Gardon and Cobonpue, 1961). This nonuniformity leads to an objectionable mottled appearance of tempered glass (Fig. 27), which may be avoided, or at least reduced, by reciprocating or orbitally oscillating the quench head in its own plane (St. Gobain, 1932; Reis, 1933).

An improved, later version of a quench head employs larger jets, operated at lower air pressures, typically 0.02–0.1 atm. This arrangement produces inherently less nonuniform heat transfer coefficients, although the jets still need to be oscillated relative to the glass plate in order to produce acceptably uniform temper. The lower air pressures also make it somewhat easier to avoid flutter of the glass between the quench heads. This low pressure air can no longer be adequately distributed by a grid of pipes. It requires quench heads in the form of plenum chambers from which issue arrays of nozzles, typically 1 cm in diameter and 10 cm long. The length of the nozzles is in large part determined by the need to provide space for the spent air to escape sideways between the glass and the plenum chamber.

2. Bent and Tempered Glass—Conventional Jet Cooling

With the increasing popularity of tempered glass, a demand also arose for curved tempered glass products. In Europe glass was at first formed in the vertical position by pressing it between a solid male mold, the surface of which was covered with many layers of fiberglass, and the rim only of the female mold. This arrangement minimized heat losses from the glass during pressing, so that it became possible for a bending station to be interposed between the exit of the furnace and the quench head. A further advantage of using a rim mold instead of a solid female die was that it could easily be articulated. Thus even relatively deeply curved parts could be formed by this method of press-bending.

To avoid multiple indexing of the glass, a later American invention (Baker, 1966) envisaged pressing the glass between two solid dies, each of which also constitutes a plenum chamber for quenching the glass through small perforations in the fiberglass-covered die faces. Upon exiting from the furnace, the glass is first pressed to shape. Then, as the dies open, quenching air is admitted through the holes in the die faces. To even out heat transfer rates, the whole die–plenum combination is reciprocated vertically, i.e., parallel to the axis of the cyclindrical bends produced by this process. In pressing, the presence of holes in the die surfaces produces barely perceptible deformations in the glass, and, for relatively small parts at least, glass can be tempered by air jets issuing not from nozzles but from holes in a perforated plate, i.e., without any specific provision being made for the escape of spent air.

In the meantime, a system for bending glass horizontally had also evolved. In this, flat glass plates are loaded on horizontal, perimeter molds, which are sent through a bending furnace along with the glass. Sagging is produced by gravity acting on the heat-softened glass, and it is controlled by careful attention to temperature and residence time. The bent glass must emerge from these furnaces when it has just the right shape and also the right temperature for quenching. With articulated molds, very deep bends and compound curved shapes can be produced (McKelvey, 1958). In such horizontal bending-and-tempering lines, the intermittent motion of vertically hung glass between furnace, pressing station, and quench head is avoided: instead, the glass moves continuously through a quenching zone. In turn, this also obviates the need to oscillate the nozzles. A common form of continuous quench head utilizes nozzles in the form of slots so oriented that the moving glass experiences a substantially uniform average rate of heat transfer over its surface (Bamford *et al.*, 1953). Glass tempered by this means is therefore less subject to objectionable optical manifestations of nonuniform tempering.

Impinging air jets produce some of the highest heat transfer coefficients attainable with gaseous heat transfer media. While local heat transfer rates are not uniform, they can readily be evened out by movement of the jets relative to the glass. Flow normal to the cooled surface is thus preferable to parallel flow, with its lower heat transfer rates that are also unavoidably nonuniform in the flow direction. While the historical development of various air quenching systems was entirely empirical, the heat transfer data now exist to design them on the basis of well-established correlations (Gardon and Cobonpue, 1961; Gardon and Akfirat, 1965, 1966).

3. Air Float Tempering

Air float tempering represents a radically different method of bending and tempering glass. It has long been known that a rigid, i.e., cold glass plate can be floated on a perforated plate supplied with air from below. A cushion of air is thus created, which is renewed from below and escapes sideways between the glass and the perforated plate. Necessarily, the pressure under central regions of the plate is higher than that under the edges, and if the glass were to be heated and softened, this nonuniform supporting pressure would tend to blow it into a bubble. To avoid this, McMaster and Nitschke (1967) replaced the perforated plate by a thicker ceramic plate with two sets of holes, as shown in Fig. 24. Through one set of holes, air to support the glass is supplied from a plenum chamber. The other set of holes, connected to exhaust manifolds within the ceramic plate, provides for the escape of "spent air" within a short distance of the supply hole. In this way the lateral flow of spent air over longer distances was avoided and, with that, the buildup of higher pressures under central regions of the glass.

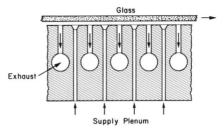

FIG. 24. Glass support block for a gas hearth (McMaster and Nitschke, 1967).

This system permitted glass plates to be "floated" into and through a furnace to heat it for tempering. By appropriately contouring the supporting
surface of such a "gas hearth," it becomes relatively easy also to bend the
glass into cylindrically curved plates as it travels forward, and to do so with
a high degree of precision. This precision derives from the fact that the
softened glass will bend in such a manner that the supporting air cushion
has the same thickness everywhere.

With the glass thus shaped and uniformly preheated, the next step is to
quench it in a similar apparatus in which the air film serves not only to
support the glass but also to cool it. One embodiment of such an apparatus
is shown in Fig. 25 (Misson, 1965). In this the glass plate is supported on
an array of closely spaced square "modules" M. Air is admitted to each of
these through a restriction O. The purpose of this restriction is to limit the

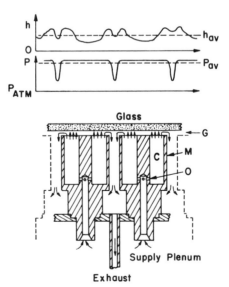

FIG. 25. Modules for supporting and quenching glass on an air film (Misson, 1965).

loss of compressed air through the module while it is not covered by glass. When the module is covered by glass, the air pressure in the chamber C builds up to the point where it supports the weight of the glass at such a distance from the module as to allow the supporting air to escape through the gap G between them. With the opening of the restricting orifice considerably smaller than that of the gap between the glass and the module, the volume of air delivered through a module is substantially constant, and as a result, the operation of this supporting device is very stable. Since air flow within the chamber is negligible, the support pressure (P) afforded the hot glass is practically uniform over the extent of the module. The only region of the hot glass not directly supported is that above the narrow gap between modules. Effects of this gap are easily ironed out by the forward movement of the glass and by the alignment of modules at a slight angle to the travel of the glass, such that no part of the glass remains unsupported for any length of time. Figure 25 also shows the spatial variation of local heat transfer coefficients h. These are governed not so much by impingement of the air, which occurs at negligible velocity, but rather by the accelerating air flow in the region where the supporting air escapes from the module. Here the air forms what has been termed a "wall jet," and heat transfer rates are high because a new boundary layer is forming between the glass and the escaping air stream. With such modules supporting the glass and cooling it from below, the glass may be cooled from above by a more or less symmetrical arrangement of modules, or with an array of nozzles designed to produce an equal heat transfer rate. Such arrangements can produce heat transfer coefficients of the order 0.02 cal/cm² °C sec, and air float quenching units have successfully been used to temper glass as thin as 3 mm. Most of the cylindrically curved tempered glass used in side windows of American vehicles is now produced on equipment of this type.

4. Quenching in Liquids and by Solid Contact

Prompted largely by the desire for ever-thinner tempered glass, increasing attention is again being given to quenching in liquids and by contact with solid platens, both of which are capable of producing the very high heat transfer rates required.

In one such process, developed in Britain (Melling et al., 1975), glass plates are quenched by being vertically dipped in liquids that include oils, molten salts, and low melting metals such as tin. Quenching in the liquid is preceded by a short exposure of the hot glass to intense gaseous cooling. The stated purpose of this is to make the surfaces rigid enough for submersion in the liquid without affecting the higher interior temperature of the glass. This process has been developed to the pilot-plant stage, and it is claimed that it can impart full temper to glass as thin as 0.75 mm.

The subject of liquid quenching has also received extensive attention in the Russian literature, and the enhanced ratios of surface compression to midplane tension obtainable in this manner have already been noted in the context of Fig. 9 (Boguslavskii *et al.,* 1964). Several workers have combined liquid quenching with chemical treatments designed to enhance the strength of the glass by etching, to remove preexisting surface flaws; by deposition of a silica-rich layer, which constitutes a form of chemical tempering, since the silica layer is likely to be further compressed by subsequent temperature equalization; or by enhancing the lubricity of the surface to protect it against future abrasion. For this a variety of liquid organic silicon compounds have been investigated (Vitman *et al.,* 1962; Boguslavskii, 1964).

As to tempering by contact with solid platens, today's glass technology and economics—unlike those of Siemens' day—require a system that preserves the surface quality of the glass plate produced in a prior operation. This means that the glass must be quenched from a lower temperature and the pressure between the glass and the quenching platens must be limited. The first of these requirements immediately raises the problem of chill cracks, made more acute by the fact that the thermal conductance between glass and metal platens is very high and also a very sensitive function of contact pressure. What makes matters worse is that abstracting heat from the glass sets up temperature gradients in the platens that tend to warp them into a convex configuration. Thus, no matter how good and uniform the contact between the platens and the glass is initially, in use the intimacy of contact is not maintained over the full extent of the glass.

The problem of chill cracks was addressed already by Siemens (1877, 1878), when he suggested preheating the platens or interposing a wire gauze between them and the glass. The use of fiberglass cloth is a contemporary version of the latter idea, which is also far less damaging to the glass surface (Ryan, 1951; Akeyoshi and Kanai, 1965). Others suggested a controlled air gap of from 0.1 to 0.4 mm between the glass and the platens (Long, 1965a) and textured metal surfaces to achieve the same effect in a more uniform and more readily controllable manner (Lehr *et al.,* 1968). Using such techniques, small specimens of very thin glass can, indeed, be tempered. For larger glass plates, thermal warping of the platens becomes the principal problem. This was recognized by Owen (1933) when he suggested the use of a flexible diaphragm to separate the glass from a liquid coolant, and cutting grooves into rigid platens to reduce their tendency to warp. Carrying this idea one step further, Nedelec (1968) envisaged cooling the glass between a rigid platen on one side and an array of individually loaded and cooled segments on the other. To get away from the marked photoelastic stress pattern produced by such discontinuities in the cooling system,

Akfirat and Gardon (1972) used a sheet of metal-filled, temperature-resistant elastomer as a covering over a rigid, water-cooled platen to give it some compliance, while also maintaining an adequate thermal conductance. To the author's knowledge, none of these schemes has yet reached the stage of commercial utilization.

C. OTHER TEMPERED GLASS PRODUCTS

While air quenching is still, and is likely to remain, the dominant method for producing automotive and architectural tempered glass, liquid quenching is used for the production of a variety of specialty glass items. Notable among these are glass piping and fittings made for heat exchangers and other chemical plant applications. These have long been made of low-expansion borosilicate glass to enhance their thermal shock resistance. By quenching in liquids, usually molten salts, these low expansion glasses can also be tempered, so that an enhanced mechanical strength is added to their intrinsically greater thermal shock resistance (Boguslavskii and Pukhlik, 1963). Liquid quenching is also used to temper the windows of spacecraft. These are made of an aluminosilicate glass, used both for its low expansion coefficient and high strain point that assures the retention of temper stresses at operating temperatures up to 700°C.

Since 1972, Federal regulations have required all glass spectacle lenses to be tempered or, at least, heat-strengthened. Both liquid and air tempering are used in the ophthalmic industry, in addition to chemical tempering.

Thermal tempering is not used in the glass container industry, mainly because of the difficulty of exposing interior surfaces to quenching air, and because liquid tempering, in the manner of de la Bastie, is impractical on the scale of current operations.

VII. Standards and Measurement Methods

A. STANDARDS FOR TEMPERED GLASS PRODUCTS

As tempered glass has come to play a progressively greater role in daily life, industrial and governmental standards have evolved to ensure its acceptability. In recent years, increased emphasis on safety has boosted regulatory activities. The standards specify minimum requirements and also the test methods by which these are to be ascertained. Thus the U.S. standards for automotive and architectural tempered glass (ANSI, 1973, 1975) specify minimum strengths under impact and the size of fragments, if breakage occurs. Tempered automotive glass, for example, must withstand the impact of a $\frac{1}{2}$-lb steel ball and an 11-lb shot-filled bag, both dropped from specified minimum heights. In the ball drop test 10 out of 12 specimens must remain unbroken. The drop height is then increased progressively until all specimens are broken. For the glass to be acceptable, none of the fragments may weigh more than 0.15 oz (4.3 g). It is a shortcoming of these tests that they are performed not on tempered articles of commerce, but on 12 in. square flat test specimens that are to be processed along with the commercial product. On one hand, it is difficult to ensure that a small, flat test specimen receives exactly the same heat treatment as, for example, a large, curved window made on a dedicated production line. On the other hand, the mass and stiffness of the two objects may also be different enough to cause them to respond differently to a given impulsive load. This is especially true for the now progressively more popular thinner tempered glass products.

The German standards for automotive glazing allow the use of tempered glass of a given thickness only after end products as well as flat specimens covering a range of sizes have passed the requisite impact tests. These standards also specify *ranges* of acceptable particle counts per unit area— rather than a maximum weight of particle—disallowing both excessively fine and excessively coarse fragmentation, but they allow some larger particles in the "vision zone" of windshields (Section V.C) in return for enhanced visibility.

It is noteworthy that these standards for tempered glass make no reference to temper stresses and that the use-oriented tests they prescribe are all destructive. The reasons for this are that nondestructive measurement of temper in large plates is not easy, none of the methods available has gained universal acceptance, and last but not least, measurement of local temper stresses in one or more regions of a plate does not conclusively ensure that the whole plate has been adequately strengthened.

B. Measurement of Temper Stresses

Temper stresses, being internally balanced stresses, are not readily amenable to *mechanical* measurement techniques, and such techniques are necessarily destructive. One great class of nondestructive test methods rests on the fact that glass is stress-optically active, so that in certain situations temper stresses can be measured *photoelastically*.

1. Destructive Tests

Apart from the use-oriented tests specified in various standards, two other destructive tests merit mention for the reason that they serve to measure temper stresses as such.

In a method due to Davidenkov and Shevandin (1939), one surface of a suitably small and flat tempered specimen is progressively etched or ground away. Removal of part of the compressive layer from only one side causes the plate to bow into a spherical shell; and, knowing the thickness of glass removed, stresses in the removed layer can be deduced from the resulting curvature of the remainder. This is a rather laborious procedure, but may well be the most accurate method for determining ordinary temper stresses in the surface of small test specimens.

The other method addresses itself primarily to the measurement of frozen-in membrane stresses in glass that is not so highly tempered as to dice when broken or cut. One such procedure consists of attaching small strain gauges to the glass in its internally stressed state, and then releasing these stresses by cutting the glass all around each strain gauge. The strain registered is then equal and opposite to the strain released by cutting the glass. Resnick and Mould (1951) coupled this approach with a polariscopic examination of the fragment, while Hubers (1964) made photoelastic measurements in the vicinity of a hole drilled into the glass to relieve local stresses.

2. Photoelastic Measurement of Temper Stresses

a. General. Temper stresses being internally balanced, photoelastic techniques are the only nondestructive means to measure them and the most important ones for research and, to some extent, for process control also. The difficulty with them is that they are not readily applicable to all tempered products and that, for certain geometries and stress systems, they may be hard to interpret.

Photoelastic measurement of temper stresses rests on the fact that most glasses are stress-optically active, i.e., they become birefringent under the action of stress. The birefringence, or optical retardation per unit path length of the measuring light beam, is proportional to the local difference of principal stresses, averaged—where appropriate—along the path. This birefringence may be gauged visually by the interference colors it produces, or it may be measured quantitatively. The basic apparatus for such observations is the polariscope, consisting essentially of a polarizer and analyzer in tandem (Coker and Filon, 1957; Frocht, 1941). Depending on the application, optical elements for imaging may also be involved, and compensators, such as a Babinet, for quantitative measurements. A number of automated or recording photoelastic stress meters have been described, using Babinet compensators for relatively large measuring ranges (Gardon *et al.*, 1966) or Sénarmont's method for very precisely measuring retardations of less than half a fringe order (Guillemet, 1965).

Interpretation of stress-optical data on tempered glass can often be simplified by choosing a location for measurement where one of the principal stresses is zero, so that the observed birefringence may be interpreted as being directly proportional to the other principal stress. Thus, to observe the distribution of ordinary temper stresses through the thickness of a glass plate, it must be viewed "edgewise." This technique is therefore clearly applicable only to relatively small, flat specimens, although it has also been used to examine larger pieces across corners, using 45° prisms to lead the light through them. In looking through a specimen edgewise (see Fig. 14), say along the z axis, with light polarized at 45° to the x and y axes and with the analyzer set at 90° to the polarizer, the polariscope measures σ_y, since $\sigma_x = 0$. Furthermore, it measures σ_y averaged along the length of the light path. Bartenev (1949) has shown that a specimen in the form of a strip having a width Z greater than four times its thickness yields a stress distribution independent of specimen size, i.e., the measured $\overline{\sigma_y}$ will be essentially unaffected by edge stresses in the specimen.

Figure 26 is a photograph of the fringe pattern in such a specimen. The upper part of the figure shows the fringes produced between crossed polaroids by the specimen alone. Except at its end, the fringes are parallel to the surfaces of the glass. They are lines of constant optical retardation and hence of constant stress. For example, the two black, zeroth-order fringes at 0.21 of the glass thickness from the surfaces represent the neutral planes in which temper stresses are zero. The lower part shows the fringes produced by the addition of a Babinet compensator in the optical path, with its fringes oriented perpendicularly to those in the glass. The zeroth-order fringe may now be recognized as a "graph" of the distribution of ordinary temper stresses across the thickness of the glass (see Fig. 1).

"Platewise" viewing a tempered plate makes the polariscope sensitive to $\overline{\sigma_y} - \sigma_z$, averaged through the plate thickness. Since ordinary temper stresses are always balanced through the thickness, this examination can, at most, yield differences of principal membrane stresses. In uniformly tempered glass membrane stresses are zero, and no birefringence is observed upon such "platewise" examination, except near the edges. There only stresses parallel to the edges can exist, and with the normal stress equal to zero, these edge stresses are easily measured and interpreted. Thus, placing a Babinet compensator over the edge, as shown in Fig. 27, allows one to visualize the distribution of edge stresses. Away from the edge, the rest of the figure shows a random pattern of black fringes, which fade into gray or white, but no colors. This is due to spatially varying membrane stresses, varying by less than $\frac{1}{2}$ fringe order, that were produced in the glass by slight nonuniformities in cooling, in the present case by impinging jets. This is the pattern that can also be observed under certain natural lighting conditions, i.e., with only partially polarized light. Its contrast is then much attenuated, giving rise, simply, to a somewhat mottled appearance of tempered automobile windows, for example. The introduction of larger, deliberate variations in membrane stresses can produce a birefringence in excess of one fringe order even in relatively thin glass.

FIG. 26. Edgewise view of a tempered strip of plate glass in a polariscope. In the lower part of the figure, a Babinet compensator has been placed in series with the specimen.

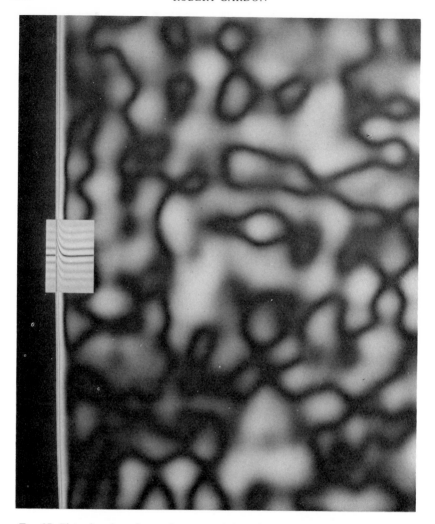

Fig. 27. Platewise view of part of a tempered glass plate in a polariscope. A Babinet compensator is used to render visible the distribution of longitudinal stresses near the edge of the plate. The mottled pattern in the interior is due to spatially varying quenching rates produced by impinging air jets. It is shown here with maximal contrast.

Figure 28 is a black-and-white photograph of the colorful photoelastic stress pattern produced by a Visurit windshield. The quantitiative interpretation of such stress patterns can be aided by the judicious use of a Babinet compensator, but it usually requires some more elaborate stress analysis (see clinopolariscope, below).

Polariscopic examination of bottles and other complex shapes is usually

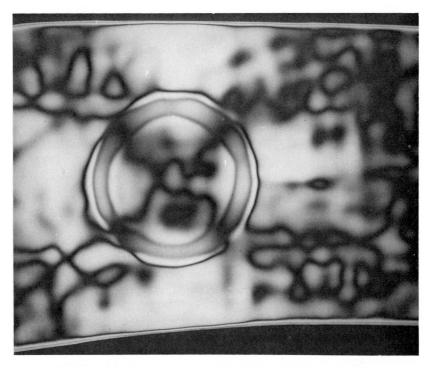

FIG. 28. Platewise view of part of a Visurit windshield in a polariscope, showing the marked birefringence due to membrane stresses in the barrier region.

only visual, experience having shown the desirable distribution of interference colors in acceptable pieces. On the other hand, quantitative measurements can sometimes also be made and justified. Thus, for example, temper stresses in glass pipes for applications in the chemical industry are routinely measured, using an immersion polariscope to avoid refraction at the glass surfaces and thus permit examination by parallel light. The resulting photoelastic stress patterns are quantitatively interpreted by the analyses of Ritland (1957) and Sutton (1958).

Photoelastic measurements using light scattered by the test object rest on the fact that scattering from a polarized light beam in the direction of a given viewing system depends on the local state of polarization of the beam, which, in turn, is a function of the stress field through which the beam has passed. In a sense, the specimen acts as its own analyzer. Used for measurements in three-dimensional objects, this technique has the virtue of revealing birefringence, and hence also stress-distributions, along the path of a light beam that can be chosen at will and yields results relatively unaffected by stresses in the remainder of the body (Weller, 1941; Frocht,

1948). Since very little light is scattered by reasonably well-homogenized glass, this technique has become attractive for use with glass only since the advent of lasers with their intense, well-collimated and monochromatic beams. Some applications to glass were demonstrated by Bateson *et al.* (1966) and Cheng (1967), and there appears to be room for further development of the scattered light technique in the glass industry.

  b. Special Photoelastic Instruments for Measurements on Tempered Glass. In addition to the above general photoelastic techniques, adaptable for various applications, a few special purpose instruments have been developed specifically for stress measurements on tempered glass, two of them for the measurement of surface compressive stresses.

  The first of these, the DSR, short for "differential surface refractometer" (see Guillemet and Acloque, 1962; Ansevin, 1965), is based on a technique first proposed by Orowan (1954). It measures the difference between the refractive indices of the surface layer of glass for two beams of light, polarized, respectively, parallel and perpendicular to the surface. It does this by measuring the difference between the corresponding critical angles for total internal reflection at the surface of a prism contacting the test piece and having a slightly higher refractive index. This differential approach allows the measurement of rather small index differences, and it eliminates the effect of any composition-related index variations in the glass. The measured difference of indices is thus entirely due to, and proportional to, the stress in the surface, normal to the plane of incidence.

  The second instrument, the "Epibiascope," measures the stress-induced optical retardation of a polarized light beam propagated *in the surface* of the glass (Guillemet and Acloque, 1960, 1962). This instrument uses a contacting prism to refract polarized light at the critical angle into the surface of the glass. As the light propagates in the surface, it is continuously attenuated by refraction back into the prism. The prism is designed to keep this twice-refracted light apart from a larger, specularly reflected component. The farther the light travels in the surface, the larger its optical retardation, an effect used to rotate the fringes produced by a Babinet compensator placed in the emergent beam. The retardation is, of course, also proportional to the stress in the surface, normal to the beam; and this is measured by measuring the rotation of the fringes.

  The intensity of light that can be refracted into and out of a glass surface is relatively small. Fortunately for the operation of the Epibiascope, this intensity is enhanced by index gradients in the tempered glass that cause a miragelike bending of light beams that travel near and nearly parallel to the surface. Similarly, index gradients near the "tin-side" of float glass also make the Epibiascope easier to use there. This "mirage effect" was put to further use in the Stratorefractometer (Guillemet, 1968), with which stress profiles and surface refractive indices in chemically tempered glass can also be measured.

The differential surface refractometer and the Epibiascope both measure the surface compression in tempered glass. They can both distinguish stresses that vary in different directions, but they cannot reveal to what extent the surface stresses measured are ordinary temper stresses and to what extent membrane stresses.

The "Clinopolariscope" developed by Acloque (1955, 1957) uses polarized light obliquely incident on the glass surface to measure the magnitude of any two mutually perpendicular components of the membrane stress at a point—not only the difference between them, which is all that can be done with normal incidence. By rotation of the instrument about its axis, the direction of the principal membrane stresses can also be found, so that, in principle at least, the stress distribution across the entire plate can be characterized.

All these instruments employ special prisms that must contact the glass to be tested through an index-matching fluid, and their alignment is rather critical. Some effort is therefore required to obtain the desired fringes in the eyepiece. Once obtained, reading them is quite straightforward, as is their interpretation. However, depending on the nature and curvature of the glass surface, readable fringes cannot always be obtained, and searching for them can become fatiguing to the eyes. As a result, the DSR, Epibiascope, and Clinopolariscope, all of which are now commercially available, tend to be used in research and for troubleshooting rather than in production. The safety of tempered glass products is therefore likely to continue to be gauged by occasional destructive tests, and a nondestructive method of quality control, applicable to 100% of production, remains an as yet unfulfilled need of the glass tempering industry.

ACKNOWLEDGMENTS

Acknowledgment is made to Dr. Paul Acloque, former Director of Fundamental Research of St. Gobain, whose work and writings greatly stimulated my early interest in the subject of tempered glass. I am indebted to Dr. O. S. Narayanaswamy, a long-time collaborator, for the insights he contributed to the overview of mathematical theories of tempering.

References

Acloque, P. (1950). *Verres Réfract.* **4**, 10–19.
Acloque, P. (1951). *Verres Réfract.* **5**, 247–260.
Acloque, P. (1955). *Verres Réfract.* **9**, 3–12.
Acloque, P. (1956). *Proc. Int. Congr. Glass, 4th,* Paper No. V-6, pp. 279–291. Chaix, Paris.
Acloque, P. (1957). *Proc. Int. Congr. Appl. Mech., 9th,* Vol. 8, pp. 415–418. Univ. of Bruxelles.
Acloque, P. (1961). *J. Am. Ceram. Soc.* **44**, 364–473.

Acloque, P. (1966). U.S. Patent 3,251,670.

Acloque, P., and Guillemet, C. (1962). In "Advances in Glass Technology" (*Proc. Int. Congr. Glass, 6th*) (F. R. Matson and G. E. Rindone, eds.), Part 2, pp. 95–106. Plenum Press, New York.

Acloque, P., and Pèyches, I. (1954). *Verres Réfract.* **8,** 69–73.

Adams, L. H., and Williamson, E. D. (1920). *J. Franklin Inst.* **190,** 597–631, 835–870.

Aggarwala, B. D., and Saibel, E. (1961). *Phys. Chem. Glasses* **2,** 137–140.

Akeyoshi, K., and Kanai, E. (1965). *Proc. Int. Congr. Glass, 7th,* Paper No. 80. Institut National du Verre, Charleroi, Belgique.

Akfirat, J. C., and Gardon, R. (1972). U.S. Patent 3,694,182.

Ansevin, R. W. (1965). *ISA Trans.* **4,** 339–43.

ANSI (1973). American National Standards Institute, Standard Z26.1.

ANSI (1975). American National Standards Institute, Standard Z97.1.

Baker, H. W. (1967). British Patent 1,092,651.

Baker, R. N. (1966). U.S. Patent 3,279,906.

Bamford, W. P., Jendrisak, J. E., and White, G. (1953). U.S. Patent 2,646,647.

Barsom, J. M. (1968). *J. Am. Ceram. Soc.* **51,** 75–78.

Bartenev, G. M. (1948a). *Dokl. Akad. Nauk SSSR* **60,** 257.

Bartenev, G. M. (1948b). *Zh. Tekh. Fiz.* **18,** 383–388.

Bartenev, G. M. (1949). *Zh. Tekh. Fiz.* **19,** 1423–1433.

Bartenev, G. M., and Rozanova, V. I. (1952). *Steklo Keram.* **9** (10).

de la Bastie, R. (1875). *Glashütte* No. 30, 149–150.

Bateson, S., Hunt, J. W., Dalby, D. A., and Sinha, N. (1966). *Bull. Am. Ceram. Soc.* **45,** 193–198.

Bertrand, P. L., and Acloque, P. H. (1964). U.S. Patent 3,149,945.

Boguslavskii, I. A. (1964). *Steklo Keram.* **21**(10), 4–9 [*English transl.: Glass Ceram. USSR* **21,** 562–567].

Boguslavskii, I. A., and Pukhlik, O. I. (1963). *Steklo Keram* **20** (9), 1–5 [*English transl.: Glass Ceram. USSR* **20,** 461–465].

Boguslavskii, I. A., and Pukhlik, O. I. (1969). *Steklo Keram.* **26** (6), 19–23 [*English transl.: Glass Ceram. USSR.* **26,** 345–349.

Boguslavskii, I. A., Vitman, F. F., and Pukhlik, O. I. (1964). *Dokl. Akad. Nauk SSSR* **157** (1), 87–90 [*English transl.: Sov. Phys. Dokl.* **9,** 587–589 (1965)].

Bunn, M. J. (1967). British Patent 1,212,208.

Carslaw, H. S., and Jaeger, J. C. (1959) In "Conduction of Heat in Solids." Oxford Univ. Press, London and New York.

Carson, F. J., Ferguson, C. W., and Ritter, G. F. (1968). U.S. Patent 3,396,000.

Chan, A. D. H., and Lambert, E. R. (1959). U.S. Patent 2,910,807.

Chan, A. D. H., and Lambert, E. R. (1963). U.S. Patent 3,081,209.

Cheng, Y. F. (1967). *Strain* **3** (2), 17–22.

Coker, E. G., and Filon, L. N. G. (1957). "Treatise on Photoelasticity," 2nd ed. Cambridge Univ. Press, London and New York.

Condon, E. U. (1954). *Am. J. Phys.* **22,** 132–142.

Davidenkov, N. N., and Shevandin, E. M. (1939). *Zh. Tekh. Fiz.* **9,** 1116.

Dennison, B. J., and Rigby, R. R. Jr., (1970). U.S. Patent 3,497,340.

Duncan, G. S. (1960). "Bibliography of Glass (from Earliest Records to 1940)." Society of Glass Tecnology, Sheffield.

Ermlich, J. R. (1965). U.S. Patent 3,169,900.

Fayet, A., Guillemet, C., and Acloque, P. (1968). *Proc. Int. Congr. High Speed Photogr., 8th* (N. R. Nilsson and L. Högberg, eds.), pp. 433–434. Wiley, New York.

Frocht, M. M. (1941). "Photoelasticity," Vol. 1. Wiley, New York.
Frocht, M. M. (1948). "Photoelasticity," Vol. 2. Wiley, New York.
Gardon, R. (1958). *J. Am. Cerm. Soc.* **41**, 200–209.
Gardon, R. (1965a). *Proc. Int. Congr. Glass, 7th* Paper No. 79. Institut National du Verre, Charleroi, Belgique.
Gardon, R. (1965b). Unpublished work.
Gardon, R. (1978). *J. Am. Ceram. Soc.* **61**, 143–146.
Gardon, R. and Akfirat, J. C. (1965). *Int. J. Heat Mass Transfer* **8**, 1261–1272.
Gardon, R., and Akfirat, J. C. (1966). *J. Heat Transfer* **88C** 101–108.
Gardon, R., and Cobonpue, J. (1961). *In* "International Developments in Heat Transfer" *(Proc. Int. Heat Transfer Conf., 2nd)* pp. 454–460. American Society of Mechanical Engineers, New York.
Gardon, R., and Narayanaswamy, O. S. (1970). *J. Am. Ceram. Soc.* **53**, 380–385.
Gardon, R., Bayma, R. W., and Warnick, A. (1966). *Exp. Mech.* **6**, 567–570.
Goldstein, M. (1964). *In* "Modern Aspects of the Vitreous State" (J. D. Mackenzie, ed.), Vol. III, pp. 90–125. Butterworths, London.
Guillemet, C. (1965). *Rev. Fr. Mécan.* No. 17, 45–53.
Guillemet, C. (1968). Thesis, Univ. of Paris, Jouve, Paris.
Guillemet, C., and Acloque, P. (1960). *In* "Colloque sur la Nature des Surfaces Vitreuses Polies," pp. 121–134. Union Scientifique Continentale du Verre, Charleroi, Belgique.
Guillemet, C., and Acloque, P. (1962). *Rev. Fr. Mécan.* No. 2/3, 157–163.
Haggerty, J. S., and Cooper, Jr., A. R. (1965). *In* "Physics of Non-crystalline Solids" (J. A. Prins, ed.), pp. 436–443. Wiley, New York.
Henrivaux, M. J. (1883). "Le Verre et le Cristal," pp. 39–55. Dunod, Paris.
Hubers, H. J. (1964). *Glass Technol.* **5**, 157–163.
Indenbom, V. L. (1954). *Zh. Tekh. Fiz.* **24**, 925–928.
Indenbom, V. L., and Vidro, L. I. (1964). *Fiz. Tverd. Tela* **6**, 992–1000 [*English transl.: Sov. Phys. Solid State* **6**, 767–772].
Jochim, F. (1965). U.S. Patent 3,186,815.
Kalman, P. (1959). *Silic. Ind.* **24**, 409–418 (in English).
Kalman, P. (1960). *J. Am. Soc.* **43**, 313–325.
Kerper, M. J., and Scuderi, T. G. (1966). *J. Am. Ceram. Soc.* **49**, 613–618.
Kitaigorodskii, I. I., and Indenbom, V. L. (1956). *Dokl. Akad. Nauk SSSR* **108**, 843–845.
Kurkjian, C. R. (1963). *Phys. Chem. Glasses* **4**, 128–136.
Lee, E. H., and Rogers, T. G. (1963). *J. Appl. Mech.* **30**, 127–133.
Lee, E. H., Rogers, T. G., and Woo, T. C. (1965). *J. Am. Ceram. Soc.* **48**, 480–487.
Lehr, G. J., Oelke, W. W., O'Connell, T. B., and Badger, A. E. (1968). U.S. Patent 3,419,371.
Long, B. (1941). U.S. Patent 2,244,715.
Long, B. (1958). U.S. Patent 2,866,299.
Long, B. (1965a). U.S. Patent 3,174,839.
Long, B. (1965b). U.S. Patent 3,174,840.
Macedo, P. B., and Napolitano, A. (1967). *J. Res. Natl. Bur. Std. Sect. A* **71**, 231–238.
McKelvey, H. E. (1958). U.S. Patent 2,827,738.
McMaster, H. A., and Nitschke, N. C. (1967). U.S. Patent 3,332,759.
Melling, R., Wright, D. C., and Pickup, J. (1975). U.S. Patent 3,890,128.
Misson, G. W. (1965). U.S. Patent 3,223,500.
Morland, L. W., and Lee, E. H. (1960). *Trans. Soc. Rheol.* **4**, 233–263.
Muki, R., and Sternberg, E. (1961). *J. Appl. Mech.* **28**, 193–207.
Narayanaswamy, O. S. (1971). *J. Am. Ceram. Soc.* **54**, 491–498.
Narayanaswamy, O. S. (1978). *J. Am. Ceram. Soc.* **61**, 146–152.

Narayanaswamy, O. S., and Gardon, R. (1969). *J. Am. Ceram. Soc.* **52**, 554–558.

Nedelec, M. (1968). U.S. Patent 3,365,286.

Ohlberg, S. M., and Woo, T. C. (1974). *J. Non-Crystall. Solids* **14**, 280–86.

Orowan, E. (1954). Private communication.

Owen, W. (1933). U.S. Patent 1,900,582.

Reis, L. v (1933). *Z. Ver. Dtsch. Ing.* **77** (23), 1–4.

Rekhson, S. M., and Mazurin, O. V. (1974). *J. Am. Ceram. Soc.* **57**, 327–328.

Resnick, I. L., and Mould, R. E. (1951). *J. Soc. Glass Technol.* **35** [167], 487–489T.

Ritland, H. N. (1957). *J. Am. Ceram. Soc.* **40**, 153–158.

Ryan, J. D. (1951). U.S. Patent 2,560,599.

Schultze, R. (1931). *Glastech. Ber.* **9**, 267–268.

Schwarzl, F., and Staverman, A. J. (1952). *J. Appl. Phys.* **23**, 838–843.

Shabanov, A. G., Boguslavskii, I. A., Pukhlik, O. I., Gasilin, E. A., and Khalizeva, O. N. (1970). *Steklo Keram.* **27** (8), 12–16 [*English transl.: Glass Ceram. USSR* **27**, 457–461].

Siemens, F. (1877). U.S. Patent 192,537.

Siemens, F. (1878). U.S. Patent 199,583.

St. Gobain, Soc. Anon. de (1931). German Patent 541,006.

St. Gobain, Soc. Anon. de (1932). German Patent 564,701.

St. Gobain, Cie. de (1966). British Patent 1,039,791.

Sutton, P. M. (1958). *J. Am. Ceram. Soc.* **41**, 103–109.

Timoshenko, S. P., and Goodier, J. N. (1970). "Theory of Elasticity," 3rd ed. McGraw-Hill, New York.

Tool, A. Q. (1946). *J. Am. Ceram. Soc.* **29**, 240–253.

Vitman, F. F., Boguslavskii, I. A., and Pukh, V. P. (1962). *Dokl. Akad. Nauk SSSR* **145**, 85–88 [*English transl.: Sov. Phys. Dokl.* **7**, 650–52 (1963)].

Weller, R. (1941). *J. Appl. Phys.* **12**, 610–616.

Weymann, H. D. (1962). *J. Am. Ceram. Soc.* **45**, 517–522.

CHAPTER 6

Chemical Strengthening of Glass

Roger F. Bartholomew
Harmon M. Garfinkel

CORNING GLASS WORKS
SULLIVAN PARK
CORNING, NEW YORK

217

I. Introduction

Glass is one of the oldest materials manufactured by man as documented by samples many thousands of years old. However, because of its brittle nature, the use of glass has been limited to nonstructural applications.

Despite intensive studies on the strength of glass, there are still many unanswered questions. Observed strengths are only a fraction of the theoretically calculated values obtained from interatomic binding potentials, approximately 2×10^6 psi (LaCourse, 1972: Doremus, 1973). It is well known that glass becomes stronger under compression and weaker in tension. Of particular interest in the study of glass fracture is the state of the surface. Figure 1 depicts the range of observed strengths and their dependence on surface condition. The presence of submicroscopic flaws, (Griffith, 1920), has been postulated to account for premature fracture. A recent review on the subject of static fatigue in glass has been written by Adams and McMillan (1977). This complex topic is discussed at length in the Chap-

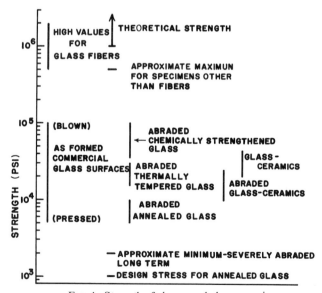

FIG. 1. Strength of glasses and glass ceramics.

ter by Freiman; for our purposes it should simply be noted that efforts to strengthen glass must take into account such surface flaws.

Two approaches have been taken to improve the strength of glass:

(a) make glass with no surface defects, or
(b) render the surface flaws inoperative.

To make high-strength glass by the first technique, special methods such as fire polishing (Proctor *et al.*, 1967) and etching (Proctor, 1962: Pavelchek and Doremus, 1974) have resulted only in temporary increases in strength, and because the surface is so susceptible to mechanical damage, the effect does not last long. Protective coating of the pristine surface by organosilicon compounds has been attempted, but with little success (Symmers *et al.*, 1962; Fletcher and Tillman, 1964).

Major increases in the strength of glass have been achieved by method (b). The simple concept underlying attempts to reduce the effects of surface flaws is to place the surface under compression; in particular, to ensure that the flaws do not extend into the tension zone. Crack propagation is slowed down so much that the tensile stress required to break the glass is equal to the strength of the untreated glass plus the additional compressive stress. Such a surface is not as susceptible to mechanical damage as an untreated surface is.

In the middle of the 17th century glass specimens known as Prince Rupert drops were shown to be very strong. These drops were made by rapidly quenching a drop of molten glass in cold water, the resulting glass has a teardrop configuration with a long tail. The thick part of the drop can be hit very hard with a hammer yet remain intact. If the tail of the drop is broken, the internal stresses are sufficiently large that violent breakage occurs; the whole object becomes fine powder.

Prince Rupert drops were the first examples, although not intentional, of increasing the strength of glass by rendering the surface flaws inoperative. Strengthening of glass today is based on two subdivisions of the technique devised to overcome the surface flaws: (a) physical (thermal) tempering, and (b) chemical tempering. A Prince Rupert drop is an example of physical tempering.

A. PHYSICAL TEMPERING

When a glass article is physically tempered, it is heated to a temperature high enough to allow relaxation of internal stresses, yet not so high a temperature as to cause deformation of the article; it is then cooled quickly (Fig. 2). The phenomena involved in this type of strengthening are discussed at length in the chapter by Gardon.

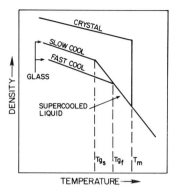

FIG. 2. Density–temperature relationship for glass.

To obtain a satisfactory stress distribution with this technique, the minimum sample thickness should be about 0.3 cm. Perhaps the most serious drawback of this method is the maximum strength, which is in the range of 25,000 psi; annealed glass has a strength of about 8000 psi.

B. CHEMICAL TEMPERING

Chemical methods of tempering have received a great deal of attention in recent years because of the greater strengths attained through these methods. Chemical tempering is based on the introduction of a compressive stress in the surface, resulting from the layer on that surface having a different composition from the underlying bulk glass. There are two techniques that can be used to arrive at such a state:

(a)   the application, at temperatures above the glass transition temperature (T_{g_f}), of a surface layer with a lower thermal expansion than the bulk glass,

(b)   the introduction of a larger ion to replace the smaller ion at temperatures below T_{g_f}. This latter process is referred to as "crowding" or "stuffing."

II. Chemical Strengthening

A. EXPANSION DIFFERENCE

A low-expansion surface can lead to strengthening using three different methods. The first is illustrated by Schott's "compound glass" first reported in 1891 (Berger, 1924), where a glass having a high thermal expansion is coated with a glass having a low thermal expansion. The drawback to such a method is that excessive differences in thermal expansion lead to sharp stress gradients at the interface between the two compositions with the danger that the coating can peel off.

A second method uses an ion-exchange process, a Li^+ for Na^+ exchange at high temperatures in a soda–lime–silica glass (Hood and Stookey, 1961). The high temperatures required tend to distort the glass article, so this method is not commercially feasible. Zijlstra and Burggraaf (1968) mentioned a similar process involving the exchange of copper ions from a cuprous halide vapor for sodium ions in a window glass composition. Strengths as high as 70,000 psi were achieved in less than 1 hr.

A third variation is possible when the low-expansion surface phase results from surface crystallization. A lithium-aluminosilicate glass can be strengthened to about 100,000 psi by formation of β-eucryptite on the surface (Olcott and Stookey, 1962). Provided the surface layers are not too thick (less than 100 μm), the crystalline phase is not very disturbing because the birefringence is weak and the effective refractive index is well matched to that of the glass. Other techniques based on this method have been described. In one technique, a soda–alumina–silica glass containing TiO_2 was subjected to Li^+ for Na^+ exchange, resulting in simultaneous crystallization (Garfinkel et al., 1962). In another, surface crystallization of a $Na_2O-Li_2O-Al_2O_3-SiO_2$ glass occurs when simultaneous exchange of Li^+ for Na^+ and diffusion of Ag^+ from a salt bath cause the formation of an intermediate layer of lithium metasilicate with a high expansion coefficient between the bulk glass and the surface layer of β-eucryptite (Cornelissen et al., 1967). More recent work is that of Partridge and McMillan (1974), who found that $ZnO-Al_2O_3-SiO_2$ glasses would crystallize from the surface during heating to give materials with flexural strengths of up to 100,000 psi. The crystal phase responsible for this strengthening is a stuffed keatite crystal having the formula $ZnO (ZnO \cdot SiO_2)$. Nucleation is enhanced by heat treatment in an oxygen atmosphere. The strengthening effect was related to both thermal expansion differences and modification of size and distribution of surface flaws by the crystal growth.

B. "STUFFING" OR ION EXCHANGE

The principle of "stuffing" was first described by Kistler (1962). He treated soda–lime–silica glass in KNO_3 melts at temperatures below 350°C and found after some time considerable compressive stress (up to 128,000 psi) at the surface. This work was followed by a report in 1964 (Nordberg et al., 1964) on K^+ for Na^+ exchange and its use for strengthening alkali–alumina–silica and alkali–zirconia–silica glasses. Since that time the greater part of the literature relating to strengthening of glass has been devoted to the ion stuffing method. The best results have been achieved with alkali–alumina–silicate glasses because it is possible to obtain a relatively high rate of exchange in these compositions with less stress relaxation (see Section V.B).

In recent years various combinations of the different methods outlined above have been attempted, e.g., a double ion-exchange treatment where a Na^+ -containing glass is first subjected to a Li^+ for Na^+ exchange above T_{g_f}, followed by a second exchange below T_{g_f}, this time by a Na^+ for Li^+ exchange (see Section VIII.A). Other types of exchanges based on similar techniques have been reported, particularly in the patent literature. The stress profiles typical of such processes tend to be complex, but controllable. Combinations of thermal tempering and ion exchange have also been tried. Such treatments will be dealt with in more detail in Section VIII.A.

Another interesting method is the combination of ion exchange and acid etching carried out by Ray *et al.* (1967). They treated glass in a bath containing an alkali-borofluoride, such as $NaBF_4$, which functions as an etchant, dissolved in molten KNO_3, which functions as the exchange media.

The advantage that most of these complex treatments have in common is that they result in a more thermally stable material of controllable breakage patterns.

III. Strength Measurements

A. MODULUS OF RUPTURE

The rupture strength of glass is usually measured by one of two different techniques; the chosen test is that which gives the most information relating to the application to which the sample is put. The two techniques are modulus of rupture (MOR) and impact testing. The former method measures the value of the maximum stress in the sample loaded to failure in a bending (or flexural) test. The latter method is a measure of the ability of the material to absorb the impact energy of a falling object.

There are many factors influencing the strength of glass, the more important factors being the severity of stress-concentration discontinuities or flaws on the surface, the test environment, and the rate of application of the stress (LaCourse, 1972). In addition, the shape of the object to be tested has to be taken into account.

Because of the presence of flaws on the surface of glass, considerable scatter is obtained in MOR data determined for as-treated samples. It was recognized early that the testing of chemically strengthened samples could be made more consistent by abrading samples such that the surface flaws were standardized. Such abrasion results in small pits in the glass that vary from 20 to 30 μm in depth. There were several methods of doing this; usually each laboratory had its own technique. Recently, the American Society for Testing and Materials (1975) has adopted standard methods of flexural testing of glass. There are basically three types of abrasion. One is a sandblast procedure using the fraction of sand passing through a No. 70 and retained in a No. 100 sieve. Alternatively, No. 30 grit silicon carbide can be used. A second procedure uses 30-grit silicon carbide and a tumble abrasion for 30 min. To withstand this technique, the compression layer must be 80–100 μm deep. After abrasion the samples are put on the test machine, such as a Tinius–Olsen, on spaced knife edges. A continuously increasing load is applied on the opposite side of the sample and in between the supporting knife edges until the sample fractures. The MOR is readily obtained from the measured load to failure, plus the geometry of the sample. These tests are carried out at room temperature in 40 ± 10% RH.

B. IMPACT TESTING

The impact test methods usually involve a stainless-steel ball of known weight dropped from increasing height until fracture occurs. This type of test is used for evaluating eyeglass lenses, which have to be able to survive a $\frac{5}{8}$-in.-diameter steel ball dropped from 50 in. (Federal Register 1971). The drop energy is merely the fracture height times the weight of the ball.

C. OPTICAL STRESS

The preceding tests were all of a destructive nature, i.e., testing to failure. However, there is a nondestructive test available for measuring the stress in glass by an optical technique. This test is based on the fact that in the absence of strain glass is isotropic; however, when stress is applied, the glass becomes birefringent, that is, doubly refracting. The ratio of birefringence to stress is called the stress-optical coefficient, or Brewster's constant. This constant is composition dependent. The measurement technique is relatively simple. Linearly polarized light when incident on a strained glass sample (between two crossed nicols) gives rise to different velocities of light in different directions and, therefore, different refractive indices. The difference in the refractive indices is called the retardation. Thus, on emerging from the sample, one component will be behind the other, resulting in elliptically polarized light. The strain is the ratio of the retardation to the index of the unstained glass. The measured optical retardation R_x in nanometers can be converted into stress through the well-known photoelasticity equation

$$(\sigma_{yy})_x - (\sigma_{xx})_x = R_x/Bd, \tag{1}$$

where B is Brewster's constant (in Brewsters or nm/cm-psi), d thickness (mm), and $(\sigma_{yy})_x$ means σ_{yy} at position x. The viewing direction is the z axis. For a more complete discussion see Morey (1954) or Rood (1972).

There are several devices and designs available for measurements. Determination can be carried out with either a polarizing microscope or a polariscope equipped with means for measuring birefrigence. Suitable instruments include the Berek or Babinet compensators (wedge prism) as well as the combination of quarterwave plate and a rotating analyzer, such as the Senarmont or Friedel compensator. Care must be taken with some methods such as the Friedel method, which is limited to compensation of one wavelength. The existence of more than one wavelength of retardation will be indicated by the presence of fringes higher than zero order. The stress as a function of distance from the surface can be readily obtained. The compressive stress at the surface of the glass plus the base strength of the unstressed sample should be approximately equal to the flexural or MOR strength of the sample. This technique differentiates very nicely between a chemically and thermally strengthened sample as shown in Fig. 3a and b.

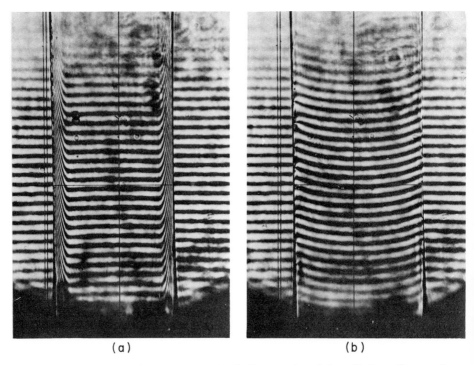

(a)   (b)

FIG. 3. Stress pattern in a sheet of (a) chemically strengthened glass; (b) thermally strengthened glass.

IV. Practical Aspects

A. MOLTEN SALTS

To carry out high-temperature ion-exchange reactions between glasses and salts, it is necessary to use ionic salts, and for the optimization of reaction rates, it has been found desirable to use molten salts as the source of the exchanging cations.

When ion "stuffing" below T_{g_f} is used, the desired temperature range is normally below 550°C for most compositions. For such temperature ranges molten nitrates have been found to be sufficiently stable (Stern, 1972) for most applications. For Na^+-Li^+ ion-exchange reactions, molten $NaNO_3$ (melting point 310°C) can be used up to approximately 450°C; above that temperature a mixed bath of $NaNO_3$ and Na_2SO_4 can be used to 525°C. Potassium nitrate (m.p. 337°C) is more thermally stable than sodium nitrate and can be used to 525°C before onset of decomposition (Bartholomew, 1966). Again in combination with sulphate, the temperature range can be extended, in the case of potassium salts, up to 600°C. The initial decomposition products of nitrate melts are nitrite and oxygen according to the reaction

$$NO_3^- \rightleftharpoons NO_2^- + \tfrac{1}{2}O_2.$$

An equilibrium is set up, which no longer holds at higher temperatures because nitrite itself decompsoes. The exact cause of the decomposition reaction is complex, because the container material influences both the rate and the path. However, below the temperature limits noted, molten nitrates can be maintained indefinitely in the liquid state without any problems. Care should be taken to exclude organic material from molten nitrate baths because of the highly oxidizing nature of the nitrates. Additions of acetates and formates to molten nitrates have been reported to lead to explosions (Kozlowski and Bartholomew, 1968). The chemistry of molten nitrates is such that normally most glass compositions can be treated without attack of the surface. Occasionally, however, the decomposition products or impurities in the bath cause surface attack varying all the way from a severe etch to a slight iridescence. To help overcome such problems, the addition of silicic acid or diatomaceous earth has been recommended (Lewek, 1968).

For higher temperatures (>600°C), combinations of salts have been reported; these include Na_2SO_4, Li_2SO_4, K_2SO_4, NaCl, and KCl. However, this is by no means an exhaustive list; the choice of the molten salt to be used is dictated by the temperature requirement as well as other considerations, e.g., thermodynamics of the exchange reaction. Lists of melting points of many salts as well as mixtures are readily available (Janz, 1967).

Care has to be taken in the selection of the container material for the molten salt. Most salts can be contained in either 309 stainless steel or Vycor brand beakers without problems. Melts containing chloride ions tend to be somewhat corrosive to the stainless steel and so are best melted in the Vycor brand beakers. In case of accidents, and to act as a heat sink, the containers should be placed in a large container surrounded by sand. With this arrangement, it is possible to maintain temperature control to $\pm 1°C$ when the salts are heated in a barrel-type furnace equipped with a temperature controller. At lower temperatures ($<500°C$), it is possible to use a mixed $KNO_3–NaNO_3$ bath as the heat sink contained in a stainless-steel vessel. Temperature control to $\pm 0.1°C$ is easily attained with such equipment.

The choice of the salt for the ion-exchange treatment depends on many factors, the main one being the exchange couple being considered. The choice of exchange conditions (time and temperature) depends upon the properties, i.e., strength, depth of layer, fracture characteristics, etc., that are required.

B. Electrical Assist

An additional driving force for ion exchange, other than the concentration gradient is that of an electrical gradient (Schulze, 1913: Spiegler and Coryell, 1952; Weber, 1965; Ohta and Hara, 1970; Plumat, 1970; Ohta, 1972, 1977). Schulze (1913) was the first to use an electric current to increase the mobility of ions in glass. Weber was the first to use an applied dc electric field to assist the migration of potassium ions into the outer surface of a bottle to strengthen glass chemically. By reversing the polarity, a compression layer can be produced on the inside of the bottle. Urnes (1973) reported the use of a constant voltage applied to sheet glass, $Na_2O–SiO_2$ and $Ni_2O–Al_2O_3–SiO_2$ glass for $Na^+–K^+$ exchange in surfaces with a source of a molten nitrate. The concentration profiles and compressive stresses were determined under different conditions of time, temperature, and voltage. These conditions depend on the amount of Na_2O in initial glass. Care must be taken not to cause resistance heating within the glass by having too high a current density as this would lead to stress release and loss of strength. For a typical soda-lime glass at 350°C, a voltage of approximately 250 V and a current of about 100 mA (this depends on the sample dimensions and electrical resistivity) for up to half an hour are sufficient to impart useful strength. The shape of the potassium concentration profile is steplike in character as shown in Fig. 4. The thickness of the potassium rich region is given by

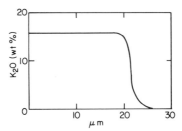

FIG. 4. Typical potassium concentration profile of electrolyzed soda–lime–silica sheet glass.

$$\Delta(t_1 t_2) = Q(t_1 t_2)/FC_0, \tag{2}$$

where $Q(t_2 t_1) = \Sigma_{t_1}^{t_2} i \, dt$ = quantity of electricity per unit area passed in time $(t_2 - t_1)$, i the current density, F the Faraday's constant, and C_0 the concentration of alkali in base glass in moles per cubic meter. Urnes showed that for thermal diffusion alone (no electrolysis) in a soda–lime composition after 60 min at 365°C the depth of penetration was 6.5 μm. However, at the same temperature, but with 195 V applied after only 26 min, the depth of penetration was 33.6 μm, over five times deeper in less than half the time. Compressive surface stresses were evaluated by measuring the optical birefringence and were reported to attain values as high as 150,000 psi.

A study of the field-assisted ion exchange of K^+ and Na^+ in soda–lime glass tubes immersed in molten KNO_3 reported that fracture was observed when a critical exchange depth was reached (Abou-el-Leil and Cooper, 1978). The major difference between this study and others (Ohta and Hara, 1970; Urnes, 1973) was a much larger exchange depth-to-glass-thickness ratio. For fracture originating on the tension side, the explanation for failure is that the tensile stress on the unexchanged side is proportional to the depth of the exchanged layer on the opposite side. Increases in this layer result in increased tensile stress until flaws on the unexchanged side undergo substantial crack growth. The crack length increases with time until the stress intensity factor reaches a critical value leading to catastropic crack propagation. Fracture from the compression side could not be satisfactorily explained.

C. PASTES

An alternative to the use of large baths of molten salts for ion exchange, which in some cases are not commercially feasible, is to use pastes. A salt containing the required alkali metal ion is mixed with an inert carrier agent in a vehicle. The resulting slurry can be adjusted to the desired viscosity for applying to the glass surface by spray coating or dip coating. The glass is allowed to air dry; several carrier agents have been found to be suitable for such a purpose. Ochre (a mixture of iron oxide and clay) shrinks when dried, and provides good binding between the paste and the glass. For applications in which Na^+ or K^+ ion exchange is required, a slurry of approximately equal parts of the nitrate and ochre mixed with two parts of water has been found to be satisfactory. The strengths of glasses treated by such a technique and by the molten salt bath process are essentially in the same range (Grego and Howell, 1973).

Pastes have been used for ion exchange of silver and copper ions in order to stain glass articles. Some recent work (Meistring *et al.*, 1976) reported that using Ag^+ containing pastes instead of the salt bath did not have any advantage. In contrast, by using a copper paste ($CuSO_4$ + clay) the rate of exchange at 560°C of copper (I) for alkali ions in a Na_2O–SiO_2 glass was faster than that for a $CuSO_4$ + Na_2SO_4, 47 : 53 mole% melt.

Although the paste technique does offer an alternative to the use of molten salt baths, it has not been applied on a commercial scale. This is because of the many steps necessary from coating to removal of the coating. Also, there has been reluctance to use pastes, because there is always a possibility that the glass surface is not completely in contact with the salt, causing pinholes in the exchange surface.

D. Vapor Phase

1. Dealkalization

Other techniques have been suggested for high-temperature ion exchange without the use of molten salt baths. One such technique is that of spraying the surface of the glass at high temperature with an aqueous solution of K_3PO_4 (Poole and Snyder, 1975) for strengthening soda–lime–silica bottles. A similar treatment has been reported using K_2CO_3 dissolved in water or an organic solvent or both (Grubb and LaDue, 1974). These exchanges must be carried out below the strain point.

The treatment of glass surfaces in atmospheres containing oxides of sulfur has been well known as a means of improving the chemical durability of glass surfaces (Keppeler, 1927; Kamita, 1930). By simply dropping a piece of sulfur in an annealing lehr, the durability will increase substantially. The reaction between sulfur oxides and glass is actually an exchange between the hydrogen ion from water in the atmosphere and sodium ions in the glass. This reaction proceeds because of the formation of a sulfate "bloom" on the surface. The reaction rate is faster if the SO_2 is oxidized to SO_3 by passing over a catalyst (Mochel *et al.*, 1966). In addition to the proton exchange, water molecules are incorporated into the glass. The source of the oxides of sulfur can be $SO_2 + O_2$, SO_3, or H_2SO_4 solution. Because this reaction is an exchange of a small ion for large ion, the treatment temperature has to be carried out above the annealing temperature. They reported that the water content of the glass is a very important parameter for the reaction rate: as the water content of the treating gas increases, the water in the surface layer of the glass increases. The net water content of the layer depends on the rate of its formation through ion exchange and the rate of loss due to thermal instability. This leads to an interesting difference between the SO_3 and SO_2 reaction that results in quite large bubbles appearing in the surface of SO_3-treated samples when water in the gas mixture is 50 wt % or more. These bubbles appeared at temperatures 30°C above the annealing point. No bubbles ever resulted from the reaction with SO_2. The modulus of rupture of soda-lime glass after SO_3 treatment did not show a significant increase for either abraded or unabraded samples in comparison with untreated samples. However, for $Na_2O-Al_2O_3-SiO_2$ glasses, nearly 300% improvement in MOR was noted with abraded specimens up to 46,100 psi. From the alkali profile in the glass, it was seen that only approximately 55% of the surface Na_2O was removed during treatment with depth of exchange of about $50\mu m$. In summary, they found that increasing the alumina content of the glass decreased the portion of alkali removed from the surface and increased the depth of reaction. The strength of cane, both with and without abrasion, also increases with alumina. Such an increase in abraded strength can at least in part be accounted for by increased depth of reaction.

Dealkalization can also be achieved using a pyrosulfate melt (Moitra *et al.*, 1971). They treated $Na_2O-MgO-Al_2O_3-SiO_2$, $Na_2O-MgO-ZrO_2-SiO_2$, and $Na_2O-CaO-SiO_2$ glasses in $K_2S_2O_7$ melts and measured dealkalization profiles that indicated depths of exchange of 20–50 μm depending on the glass, time, and temperature. The abraded MOR strength of the treated samples showed values as high as 64,000 psi in the alumina-containing glasses.

2. Miscellaneous Vapor Phase Methods

Ion exchange from the vapor phase has been tried using cuprous halides as well as lithium halides (Shonebarger, 1970, 1971). By exposing soda–lime–silica compositions above their annealing temperature to CuCl, CuBr, or CuI vapor, Shonebarger was able to strengthen the glass. To reduce the probability of the disproportionation reaction

$$2Cu^+ \rightarrow Cu^0 + Cu^{2+},$$

the glasses were melted with reducing agents present. The abraded strengths of treated rods were as high as 70,000 psi (untreated samples 22,000 psi) when treated in CuCl vapor. Less strengthening was seen with

CuI and even less with CuBr. Because the radius of the Cu^+ ion is very nearly equal to that of the Na^+ ion, it would seem that the mechanism responsible for the increase in strength arises from a difference in thermal expansion between the exchanged surface layer and the interior.

Faile and Roy (1971) reported that the high-pressure impregnation of argon into fused silica increased its flexural strength. Pentration depths of up to 60 μm were determined using electron microprobe measurements. Glass impregnated at 2 Kbar and 650°C increased the strength from 17,600 to 20,800 psi. Similar experiments with N_2 and H_2 showed no increase in strength. Argon impregnation of borosilicate and soda–lime–silica glasses did not raise the strength. The authors attributed this fact to too low a penetration of gas into the glass. Impregnation had to be carried out at temperatures below the strain points of the glasses, so that stress relaxation did not occur. At these low temperatures little diffusion of Ar could take place; only for fused silica could high-enough temperatures be used.

Plumat and Toussaint (1972) reported that a CO_2 atmosphere was beneficial for ion exchange using nitrate melts. The CO_2 gas increased the diffusion rate; unfortunately, the authors were not able to explain this phenomenon.

Before leaving the subject of single-treatment methods for strengthening glass using vapor phase methods, mention should be made of the high strengths reported on autoclaving alkali silicate compositions (some containing PbO or BaO) in steam (Stookey, 1970). A surface layer of hydrated glass produced over a core of substantial thickness of original glass leads to modulus of rupture values ranging from 80,000 to 250,000 psi. This improvement in strength is ascribed to a simultaneous polishing action and formation of a low elastic modulus protective skin. Similar increase in strength in soda–lime–silica glass films treated in steam at about 190°– 260°C has been reported (Charles, 1958). Transverse rupture strengths of 20,000–25,000 psi were obtained. In this composition the strength increase was considered to arise from a corrosion protection product layer, a few mils in thickness, which prevented surface damage during subsequent handling.

V. Ion Exchange

A. KINETICS

1. Theory

When a glass containing monovalent cations is placed into a molten salt containing another monovalent cation, ion exchange takes place. Schulze (1913) was the first to demonstrate that glass was an ion exchanger. A generalized ion-exchange reaction can be written

$$\overline{A} \text{ (glass)} + B \text{ (salt)} = \overline{B} \text{ (glass)} + A \text{ (salt)}, \tag{3}$$

where $\overline{A}$ and $\overline{B}$ are the counterions in the exchanger phase and A and B are the counterions in the liquid phase. Molten salts are required because of the temperature range needed before the cations in the glass become mobile with reference to the negatively charged oxygens of the rigid, immobile silicate network. There are several rate-controlling mechanisms possible; however, it has been clearly established that the rate at which exchange occurs is controlled by diffusion of the ions in the glass. Study of the diffusion process and the parameters, which influence the mobility of the cations, can provide insight into the mechanism and, in the case of chemical tempering of glass, lead to a better understanding of the mechanisms. In this section exchange kinetics are discussed. Self-diffusion and interdiffusion and the relationship between the two will be discussed further.

Because most of the techniques used to study exchange kinetics have been described in some detail in the literature (Garfinkel, 1972; Frischat, 1975), they are only discussed briefly here. In all the methods described in the following sections, the conditions are experimentally fixed so that it can be assumed that (1) net diffusion is unidirectional and (2) the sample extends semi-infinitely in the direction parallel to the gradients in concentration.

In binary ion exchange, the diffusing species A and B are charged, and tend, in general, to diffuse at different rates. Thus there is a tendency for one ion to move faster than the other, leading to a buildup of electrical charge. There is, however, a gradient in electrical potential along with this charge, which acts to slow down the faster ion and speed up the slower one. Despite the difference in the mobilities of the two ions, the gradient in electrical potential forces the fluxes of the two ions to be equal and opposite, thus preserving electrical neutrality.

The driving force for transport of species A or B is the negative of its gradient in total chemical potential μ. The total chemical or electrochemical potential is the sum of the gradient in activity and the gradient in electrical potential. This assertion is based upon the assumption that no other gradients or thermodynamic forces, such as gradients in pressure or gradients in temperature, are present in the system. With the additional assumption that the mobilities of the ions are the same in self-diffusion and interdiffusion, the flux of diffusing species per unit time in the x direction is (Doremus, 1964)

$$J_A = -D_A \left[\frac{\partial \overline{c}_A}{\partial x} \frac{\partial \ln \overline{a}_A}{\partial \ln \overline{c}_A} + \overline{c}_A \frac{FE}{RT} \right], \tag{4}$$

where D_A is the self-diffusion coefficient, $\overline{c}_A$ the ionic concentration, $\overline{a}_A$ the activity of species A, and E the potential gradient or electric field; a similar expression can be written for the flux J_B. Only univalent-for-univalent exchange is under consideration here. This set of flux equations is the Nernst–Planck equations.

The conditions of electrical neutrality require that $J_A = -J_B$. Since the number of negative exchange sites is constant and fixed, $\partial \bar{c}_A/\partial x = -\partial \bar{c}_B/\partial x$

$$E = \frac{RT}{F} \left[\frac{D_B - D_A}{\bar{c}_A D_A + \bar{c}_B D_B} \right] \frac{\partial \bar{c}_A}{\partial x} \frac{\partial \ln \bar{a}_A}{\partial \ln \bar{c}_A}. \tag{5}$$

Substitution of Eq. (5) into Eq. (4) gives

$$J_A = \frac{D_A D_B}{\bar{N}_A D_A + \bar{N}_B D_B} \frac{\partial \ln \bar{a}_A}{\partial \ln \bar{c}_A}, \tag{6}$$

where $\bar{N}$ is the cation fraction in the exchange; a similar equation can be written for J_B. The activity can be related to the concentration (Karreman and Eisenman, 1962) by

$$a_A = \bar{c}_A^n, \tag{7}$$

n being a constant. By comparison with Fick's first law for diffusion, since $\partial \ln \bar{a}_A/\partial \ln \bar{c}_A = n$ from Eq. (7), the interdiffusion coefficient becomes

$$\bar{D} = \frac{D_A D_B}{\bar{N}_A D_A + \bar{N}_B D_B} n. \tag{8}$$

For ideal solutions $n = 1$, so one gets the commonly used relationship between the interdiffusion coefficient and the self-diffusion coefficients of the individual cations. These equations were first applied to the interdiffusion of hydrogen and sodium ions in polysulfonate exchangers in aqueous solution (Helfferich, 1962; Helfferich and Plesset, 1958).

The temperature dependence of the interdiffusion and self-diffusion coefficients follows the Arrhenius equation

$$D = D_0 \exp(-E\ddagger/RT) \tag{9}$$

in which $E\ddagger$ is the activation energy.

If the electrical current in an exchanger is carried by a single ionic species, then the electrical conductivity of the exchanger can be related to the diffusion coefficient of this species by the Nernst–Einstein equation

$$D = \sigma RT/c(zF)^2, \tag{10}$$

where σ is the specific conductivity in $\Omega^{-1}cm^{-1}$, R is 8.314 J/deg mole, c the concentration of the diffusing species in moles/cm^3, z the ionic value, and F is Faraday or 96,500 C/equiv.

It has been found that the measured electrical conductivity of glass is less than that calculated from diffusion data (Doremus, 1962; Stern, 1966; Terai and Hayami, 1975). These results have been interpreted in terms of a model that accounts for the difference in terms of the two kinds of processes. In this model a correlation factor f, which depends on the structure of the exchanger and on the transport mechanism, is used to modify the Nernst–Einstein equation so that

$$f = D_A/D_\sigma, \tag{11}$$

where D_A is the measured self-diffusion coefficient, and D_σ the diffusion coefficient calculated from the conductivity with Eq. (10). Thus f is a measure of the efficiency of the Nernst–Einstein equation. If diffusional motion is completely random, $f = 1$. However, f is less than unity if the next jump depends upon the direction of the previous jumps, i.e., the jump direction is not entirely random. Since Compaan and Haven (1956) showed that such correlations do not arise in electrical conduction, the factor f can be attributed to the correlations in diffusion. Haven and Stevels (1957), and later Haven and Verkerk (1965), discussed the possible mechanisms and the magnitude of the correlation effects associated with these mechanisms in glass that could account for the observed relation between diffusion and ionic conductivity.

2. Experimental

Experimental data for both self-diffusion and interdiffusion kinetics can be obtained by several techniques. To obtain self-diffusion coefficients, radioactive tracers are necessary. Weight change, sectioning, and electron microprobe data have been used most extensively for interdiffusion measurements, and resistance change to a lesser content. These methods are described individually below.

a. Weight Change. By measuring the weight of the exchanger before and after the exchange, the total amount in moles of A taken up per unit surface area of glass exposed to the molten salt is

$$Q_A = \frac{\Delta w}{S(M_A - M_B)}, \tag{12}$$

where M is the gram-atomic weight, Δw the change in weight of the sample following ion exchange, and S the superficial area.

Integration of Fick's first law with constant mean integral interdiffusion coefficient D_{AB} gives

$$Q_A = 2C_{0,A} (D_{AB}t/\pi)^{1/2}, \tag{13}$$

where $C_{0,A}$ is the surface concentration per unit volume. Thus, the amount of material taken up by the exchanger is proportional to the square root of time. In order to double the amount of reaction at a given temperature, the time of treatment must be increased fourfold. If the interdiffusion coefficient is a function of concentration, then D_{AB}, determined from Eq. (13), will be some mean value over the initial and final concentration in the sample. In this case, the uptake is written

$$Q_A = \int_0^t J_A^0 \ dt = -2(\hat{D}_{AB}^0 t)^{1/2} \left(\frac{\partial c_A}{\partial w}\right)_{w=0} \tag{14}$$

where $w = x/(\hat{D}_{AB}^0 t)^{1/2}$; J_A^0 is the flux of A, and $\hat{D}_{AB}^0$ the interdiffusion coefficient, respectively, at $x = 0$. Thus, Q_A is still proportional to the square root of the time of exchange, because $(\partial c_A/\partial w)_{w=0}$ is constant for a constant profile shape irrespective of the depth of exchange (Doremus, 1964; Garfinkel, 1970). The square-root-of-time law for uptake is well established for ion exchange of glass in molten salts (Doremus, 1962, 1964; Nordberg et al., 1964; Stern, 1966; Garfinkel, 1970). Figure 5 illustrates typical weight change data.

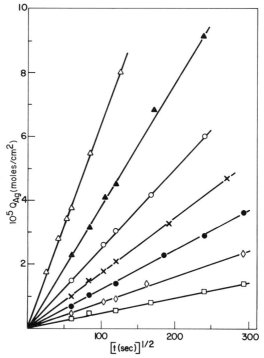

FIG. 5. Weight change as a function of $t^{1/2}$. Ag$^+$–Li$^+$ ion exchange in 11.4 mole % Li$_2$O, 16.5 mole % Al$_2$O$_3$, 71.5 mole % SiO$_2$ glass. $\triangle$, 395°C; $\blacktriangle$, 352°C; $\circ$, 320°C; $\times$, 306°C; $\bullet$, 277°C; $\diamond$, 254°C; $\square$, 228°C.

b. Sectioning. The concentration distribution can be determined by sectioning the exchanger following ion exchange. Thin, successive portions of the exchanged sample are removed with an $HF-H_2SO_4$ solution. The depth removed is calculated from the density of the glass, the mass of the section removed, and the total surface area exposed to the etchant, i.e.,

$$d = \Delta w - m_B(M_B - M_A)/\rho_0 S_0 \tag{15}$$

where Δw is the mass of the section removed, m_B the total number of moles of B in the layer, M the molecular weight, ρ_0 the density of the unexchanged sample, and S the surface area. Analysis of the etchates by flame photometry or titrimetry gives the ionic composition of the removed layer. If the bath is tagged with radiotracer, then counting the etchates or remaining sample will yield information about the diffusion profile.

For the case in which the etched solution is analyzed either by flame photometry or counting methods, the diffusion coefficient is determined from Garfinkel and Rauscher (1966)

$$A_j/A_0 = 1 - \text{erf } \xi_j \tag{16}$$

where A_j/A_0 is the ratio of the activity of the jth layer to that at the surface, erf ξ_j the error function or probability integral, and $\xi_j = x_j/2 (Dt)^{1/2}$. For the residual counting technique, i.e., the method in which the entire sample was counted before and after etching, the diffusion coefficient is obtained from

$$A_j/A_0 = \pi^{1/2} \text{ ierfc } \xi_j \tag{17}$$

where A_j/A_0 is the ratio of the activity left in the sample after removal of the jth layer to the total activity in the sample, $\xi_j = x_j/2 (Dt)^{1/2}$, and ierfc ξ_j is the first integral of the error-function complement. Equations (16) and (17) are applicable only for the case in which the diffusivity is independent of concentration; these equations are especially applicable to self-diffusion studies. Figure 6 is an example of interdiffusion coefficients measured by this method. If the interdiffusion coefficient is a function of concentration in the sample, then numerical methods like Fijita's solution (Crank, 1956) must be used.

c. Electron Microprobe. The sample, in the form of a rod, is exchanged in the molten salt. A section is removed from the center of the rod, mounted in a plastic support, and polished. The profile is then determined with the electron microprobe (Guilliemet *et al.*, 1967). Care has to be taken to minimize migration of the alkali ion due to the electron beam heating (Borom and Hanneman, 1967) and charging (Varshneya *et al.*, 1966: Vassamillet and Caldwell, 1969), which can cause serious errors. This is

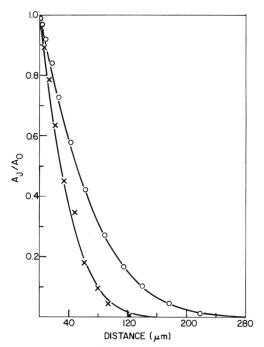

Fig. 6. Residual–concentration profiles. Na–Li$^+$ ion exchange in Corning 9608 glass ceramic. ○, 394°C, 17.2 hr; ×, 349°C, 17.7 hr.

observed when count rate varies with time. Modification of equipment such as that described by Kane and Williams (1971) for the ARL–EMX electron microprobe reduces this problem. A discussion on this problem is given in a review by Kane (1973). Inaccuracy in establishing a sharp boundary at the surface of glass with the mounting material can lead to other problems (Ish-Shalom and Winitzer, 1972; Varshneya and Milberg, 1974). Varshneya and others have discussed the effect that the electron beam has on finite width of about 2–3 μm (Varshneya and Milberg, 1974; Varshneya and Menick, 1974). The diffusion coefficient as a function of local composition is usually computed from the profile of weight percent alkali metal oxide against distance, using the Boltzmann–Montano technique (Crank, 1956).

$$D(c) = -\frac{1}{2t}\frac{dx}{dc}\int_0^c x\ dc \qquad (18)$$

where $D(c)$ is the concentration-dependent diffusion coefficient (cm^2/sec), c the concentration (gm/cm^3), t the time (sec), and x the distance (cm) from the surface. A typical concentration–distance profile determined by the microprobe is shown in Fig. 7.

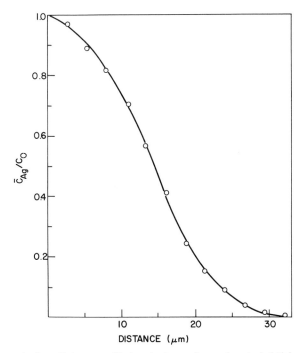

FIG. 7. Concentration–distance profile by electron microprobe. Ag^+–Li^+ ion exchange in 11.4 mole % Li_2O, 16.5 mole % Al_2O_3, 71.5 mole % SiO_2 glass; 365°C, 1 hr.

An electron beam is focused to a line oriented perpendicular to the direction of diffusion, then the sample is stepped at intervals of several micrometers in the direction of diffusion. Characteristic x-ray intensities for the element under study are collected at each step for a present probe current. The counts obtained can be converted to weight percent oxide using calibration curves obtained from glass standards. Most modern electron microprobe instruments are capable of measuring elements heavier than lithium.

d. Microindentation. A novel method of determining the concentration profile has been reported by Frischat (1970). A simple microindentation method using the Vickers hardness measurement technique was applied to a K^+–Na^+ exchanged sample of a Na_2O–Al_2O_3–SiO_2 glass. A depth of ion exchange of approximately 150 μm was obtained, and microhardness was determined after successive removal of 5–10 μm thick layers. With the assumption that the increase in microhardness was directly proportional to K^+ concentration in the glass, Frischat fitted the microhardness-penetration depth profile according to Eq. (19), which is derived from Fick's second law.

$$C_{K+} (x, t) = C_0 [1 - \text{erf}(x/2(Dt)^{1/2})].   (19)$$

He was able to fit the data using a concentration-independent interdiffusion coefficient. Other glasses of the systems Na_2O-SiO_2 and $Na_2O-Al_2O_3-SiO_2$ showed similar results. However, a $Na_2O-CaO-SiO_2$ glass did not show a simple relationship between Vicker's hardness and K^+ concentration.

e. Changes in Electrical Resistance. The change in resistance is measured in two ways. In the first method, the resistance is measured at temperature as diffusion progresses; by determining, the resistance of the unexchanged glass at temperature, the change in resistance is known as a function of time of exchange (Doremus, 1964; Garfinkel, 1970). In the second method, the resistance is measured on exchanged disks as a function of temperature to 100°C below the exchange temperature to ensure that the concentration distribution remains fixed. The specific resistivity of the unexchanged sample is also determined. With these data the resistance before and after ion exchange is extrapolated to the actual temperature of exchange to give the change in resistance (Garfinkel, 1970).

The electrical resistance of the exchanger after exchange for time t is given by

$$R_t = \frac{1}{S} \int_0^L \frac{dx}{\sigma_t},   (20)$$

where S is the superficial area of the exchanger exposed to the molten salt, L the thickness of the exchanger, and σ_t the specific conductivity of the exchanger at a point x from the glass surface. Therefore, the resistance of the exchanger at any time t of exchange is given by

$$R_t - R_0 = \frac{(D_A t)^{1/2} R_0}{L} \int_0^{L/(D_A t)^{1/2}} \frac{(\sigma_0 - \sigma_t)}{\sigma_t} dy.   (21)$$

where $y = x/(D_A t)^{1/2}$, D_A is the self-diffusion coefficient of the ion originally present in the exchanger, R_0 the resistance, and σ_0 the specific conductivity of the exchanger in its original form. Since the integral in Eq. (21) is constant as long as the depth of exchange is short compared to the thickness of the exchanger, i.e., $L >> (D_A t)^{1/2}$, the change in resistance is proportional to the square root of the time of ion exchange.

B. THERMODYNAMICS

1. Theory

From the ion-exchange reaction written in Eq. (3), an equilibrium constant K_{AB} can be defined so that

$$K_{AB} = \bar{a}_B a_A / \bar{a}_A a_B. \tag{22}$$

The $\bar{a}_i$s in Eq. (22) are the respective activities in the exchanger phase and the a_is are the respective activities in the liquid phase. The value of K_{AB} depends upon the reference state chosen to define activities. For the salt, it is convenient to use the pure material as reference state, so that

$$\lim_{N_A \to 1} \gamma_A = 1 \tag{23}$$

for component A, and similarly for component B. Here γ_A is the activity coefficient of component A. The reference state for the solid exchanger is that in which all of the exchangeable cations are of the ion in question.

Ion-exchange equilibrium is characterized by the ion-exchange isotherm, which is a graphical representation covering all experimental conditions at constant temperature. Of prime significance is the selectivity of the exchanger, i.e., the selection of one counterion in preference to the other by the ion exchanger. The equilibrium constant K_{AB} is an integral measure of selectivity.

Certain assumptions must be used to get at the experimental quantities in Eq. (3). The ratio of the activites of the ions in the exchanger is given by Rothmund and Kornfeld (1918):

$$\bar{a}_B / \bar{a}_A = (\bar{N}_B / \bar{N}_A)^n, \tag{24}$$

where $\bar{N}$ is the cation fraction. This has been referred to as n-type behavior in aqueous sytems (Karreman and Eisenman, 1962).

Most of the molten nitrate mixtures used can be characterized as regular solutions (Sundheim, 1964). Although the heat-of-mixing of these salts is a slight function of composition, the approximation is quite satisfactory. Therefore we may write

$$RT \ln \frac{\gamma_A}{\gamma_B} = A(1 - 2N_A), \tag{25}$$

where A is a constant independent of temperature. Substitution of Eqs. (24) and (25) into Eq. (22) yields

$$\log \frac{N_B}{N_A} - \frac{A}{2.303RT} (1 - 2N_A) = n \log \frac{\bar{N}_B}{\bar{N}_A} - \log K_{AB}, \tag{26}$$

which is similar to the semiempirical relation proposed by Kielland (1935) for aqueous exchangers (Garfinkel, 1968). Thus a plot of $\log(a_B/a_A)$ for the salt against $\log(\bar{N}_B/\bar{N}_A)$ for the exchanger is characterized by a slope of n if K_{AB} is constant with concentration. Typical data are shown in Figs. 8 and 9.

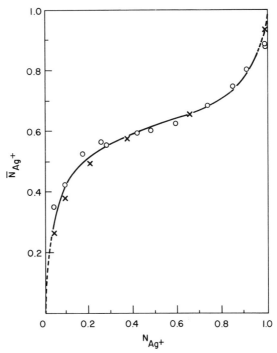

FIG. 8. Ion-exchange isotherm: ○, 300°C; ×, 451°C; Ag^+–Li^+ ion exchange in 11.4 mole % Li_2O, 16.5 mole % Al_2O_3, 71.5 mole % SiO_2 glass.

2. *Experimental*

This discussion is concerned with two direct measurements (Garfinkel, 1968) as opposed to indirect measurements involving membrane potentials.

In the first method, powdered exchanger, which passes a −270 (53 μm) mesh screen, is equilibrated for 200–500 hr, depending upon temperature, in a molten salt bath with continuous stirring. The glass powder is analyzed either by flame photometry, or by gamma counting for those instances in which the bath is tagged with radiotracer.

In the second method, the concentration–distance profile is determined on an ion-exchanged sample in the form of a rod. The sample is etched away with HF–H_2SO_4, and the etchate at different distances into the sample is analyzed either by gamma counting or by flame photometry, depending upon whether the bath is tagged with radiotracer or not. Then the concentration–distance plot is extrapolated to the surface $x = 0$ to obtain the surface concentration.

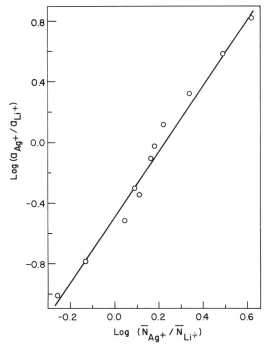

FIG. 9. Test of n-type behavior for exchange system Ag^+–Li^+ at 300°C; $n = 2.2$; $w = -2.7$ kcal mol^{-1}; in 11.4 mole % Li_2O, 16.5 mole % Al_2O_3, 71.5 mole % SiO_2 glass.

VI. Mechanical and Physical Properties of Chemically Strengthened Glass

A. ORIGIN OF STRESS

As mentioned at the beginning, glass is intrinsically a very strong material. The intrinsic strength of silica glass, based on the Si–O bond energy, is approximately 2×10^6 psi. However, the presence of Griffith flaws on the surface reduces the strength to less than 8000 psi in most cases. The exchange of a large ion for a small ion at a temperature below that at which the glass network can relax produces a concentration distribution across the surface of the glass, resulting in a stress profile. The larger volume of the incoming ion produces the compressive stress that is analogous to a temperature distribution. This analogy is obtained by substituting the linear lattice expansion coefficient for the linear thermal expansion coefficient, and the temperature by the concentration. This problem has been treated by several authors (Cooper and Krohn, 1969, Garfinkel and King, 1970; Schaeffer and Heinze, 1974; Varshneya, 1975), and they have shown stress to be related to concentration for a slab of thickness $2l$ by

$$\sigma_{zz}(x,\ t) = -\frac{BEc}{1-\nu} + \frac{BE}{2l(1-\nu)}\int_{-l}^{+l} c\ dx, \tag{27}$$

where E is Young's modulus, ν Poisson's ratio, and B the lattice dilation coefficient; B, E, and ν are assumed constant. Since

$$\frac{M_t}{M_\infty} = \frac{1}{lc_0}\int_{-l}^{+l} c\ dx, \tag{28}$$

Eq. (27) can be rewritten

$$\sigma_{zz}(x,\ t) = \frac{BE}{1-\nu}\left[c_0\frac{M_t}{M_\infty} - c(x,\ t)\right]. \tag{29}$$

For stresses at the surface,

$$\sigma_{zz}(l,\ t) = \frac{BEc_0}{1-\nu}\left(\frac{M_t}{M_\infty} - 1\right). \tag{30}$$

The normalized stress F_σ is defined by

$$\frac{\sigma_{zz}(x,\ t)}{\sigma_{zz}(l,\ t)} = F_\sigma\frac{c(x,\ t)/c_0 - M_t/M_\infty}{1 - M_t/M_\infty}. \tag{31}$$

Thus F_σ can be calculated from the experimentally determined concentration–distance profile, where M_t/M_∞ is obtained by graphical integration of Eq. (28). Figure 10 compares F_σ calculated from Eq. (31) and measured values, with the independently determined experimental value of M_t/M_∞ = 0.105, for a Li_2O–Al_2O_3–SiO_2 glass treated in molten $NaNO_3$ at 400°C for 4 hr. The thermal stress analogy overestimates the depth of the compressive layer by about 16%. This lack of agreement could result from a dependence of E and ν on position, occurrence of stress relaxation during exchange, or production of bending stresses by asymmetric stress distribution.

In the absence of stress-relaxation processes, the linear strain on the surface of a glass before and after ion exchange (100% surface exchange) would be approximately

$$1/3[(V_{exch} - V)/V], \tag{32}$$

where V is the molar volume of the unexchanged glass and V_{exch} the molar volume of the exchanged glass. Values around 5% would be predicted for glasses involving K^+–Na^+ exchange. Varshneya (1975) has used Eq. (27) to calculate the surface compression expected from exchange of K^+ for Na^+ in a $Na_2O\cdot 3SiO_2$ glass. He obtained a value of approximately 400,000 psi with B equal to 1.3×10^{-3} (wt % K_2O). These values are about three to

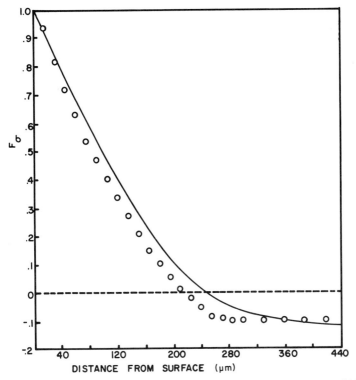

FIG. 10. Comparison of F_σ measured (circles) and calculated from Eq. (31) (solid line). 5% Li_2O, 26% Al_2O_3, 67% SiO_2, 1% TiO_2 (wt %) at 400°C, 4 hr in $NaNO_3$.

four times those measured by Cooper and Krohn (1969). By a slightly different approach, Hale (1968) arrived at a similar lack of agreement between calculated and experimental strength values. He used thermodynamic agreements, then assumed that glass is an elastic continuum and applied the "misfitting" sphere theory (Eshelby, 1956). The compressive stress P at the surface was found to be given by

$$P = (\Delta V_0/3V_0)[E/(1 - \nu)], \qquad (33)$$

where ΔV_0 is the increase in volume of a volume V_0 of glass containing 1 mole of ion A, when the A ions are replaced by B ions. Hale (1968) suggested that the glass behaved as if the size of a sodium interstice (in a melted glass) corresponded to a radius of 1.15 Å instead of 0.94 Å. This leads to a change in volume of 2.1 cm^2 $mole^{-1}$, and a surface compression of 154,700 psi, which Hale claims to be in reasonable agreement with 118,-000 psi obtained experimentally by Burggraaf (1966) with a glass containing

20.15 wt% Na_2O. In the same paper Burggraaf showed that the glass obtained by ion exchange has a considerably greater density than glass of the same composition obtained by a normal melting procedure. As a result the volume changes are less than predicted, hence the observed stresses are lower than the calculated values.

B. STRESS RELAXATION

The preceding calculations were based on the assumption that there is no stress relaxation during ion-exchange treatment of a glass. What then is the effect of time and temperature of treatment on strength? There appears to be a time at which the modulus of rupture of a glass goes through a maximum for each treatment temperature. The higher the temperature, the shorter the time to attain this maximum strength. Figure 11 illustrates this behavior for a $Na_2O-Al_2O_3-SiO_2$ glass. This can be understood better if we consider that the rate of buildup of the integral stress is proportional to the rate of ion exchange minus loss in stress due to relaxation of the glass (Acloque and Tachon, 1962, Garfinkel, 1969). Thus, we may write

$$\frac{d\sigma}{dt} = \frac{k}{t^{1/2}} - \frac{\sigma}{\tau}, \tag{34}$$

where τ is the relaxation time.

Dimensionally, τ is the viscosity divided by the bulk modulus. However, the operational significance of these quantities at the temperatures of interest for the densified surface layer produced by ion exchange is not very clear.

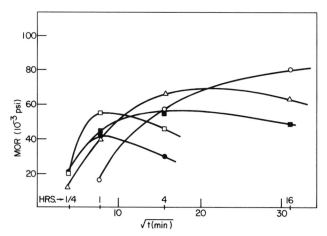

FIG. 11. Effect of temperature on the strength of potassium-exchanged 1.1 Na_2O-1 Al_2O_3- 4 SiO_2 glass. ○, 450°C; △, 500°C; ■, 525°C; □, 550°C; ●, 575°C.

At any rate, multiplying both sides of Eq. (34) by exp (t/τ) and integrating gives

$$\sigma = 2k\tau^{1/2} \exp(-x^2) \int_0^x \exp(y^2) \, dy, \tag{35}$$

where $x = (t/\tau)^{1/2}$. Equation (35) can be rewritten

$$\sigma = i(\pi\tau)^{1/2}k \exp(x^2) \operatorname{erf} x \tag{36}$$

or

$$\sigma = 2k\tau^{1/2}F(x). \tag{37}$$

The function $F(x)$ has been tabulated by Miller and Gordon (1931). It is easily verified that

$$\lim_{\tau\to\infty} \sigma = 2kt^{1/2} \tag{38}$$

and

$$\lim_{t\to\infty} \sigma = 0. \tag{39}$$

Thus, for very large relaxation times, the integral stress should increase linearly with the square root of the time of treatment as long as the sample remains infinite in extent. However, as the time of treatment becomes large with respect to the relaxation time τ, the stress will begin to drop off from this linear relationship. The integral stress should exhibit a maximum value at $t_{max} = 0.853\tau$. One would expect the temperature dependence of the relaxation time τ to be given by

$$\tau = \tau_0 \exp(E/RT). \tag{40}$$

The function $\sigma/2k$ is plotted in Fig. 12 as a function of the square root of the treatment time for different values of the relaxation time τ. Note the striking similarity between the experimental curves in Fig. 11 and the calculated curves in Fig. 12. Thus, the maximum arises because of the competition between the stress buildup as a result of the ion exchange and the stress release as a result of the thermal accommodation by the glass network. Such an accommodation mechanism, in terms of the relaxation time of the densified layer produced by ion exchange, can be used to discuss phenomenologically the difference in degree of strengthening of soda–lime–silica and soda–alumina–silica glasses.

In view of Eq. (40), a plot of log τ versus the reciprocal of the absolute temperature should yield a straight line, and the temperature coefficient of τ is given by the slope of this line. A semilogarithmic plot of τ versus the reciprocal of the absolute temperature is shown in Fig. 13 for soda–alu-

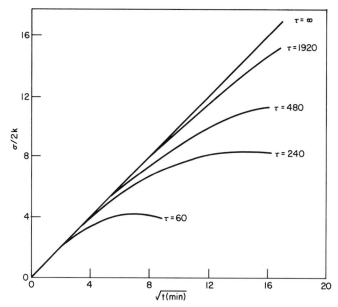

FIG. 12. Calculated effect of time of exchange and relaxation time on strength of ion exchanged glass.

mina–silica ($1.1Na_2O–Al_2O_3–4SiO_2$) and a soda–lime–silica (Corning Code 0080) glass. Since t_{max} is a function of the abrasion used, only the apparent relaxation time is plotted. In this case, 150-grit abrasion was used exclusively. The values of τ for the soda–lime–silica glass were obtained from data on Corning Code 0080 similar to those shown in Fig. 11 (Garfinkel, 1969).

The temperature coefficient of τ for the alumina glass is 32 kcal/mole. This value, which is lower than expected, is in the range of the temperature coefficient of the ion exchange itself. On the other hand, the temperature coefficient of τ for the Code 0080 glass is 58 kcal/mole. At 475°C the exchanged densified surface layer of the soda–lime–silica glass relaxes about as rapidly as the exchanged layer of the soda–alumina–silica glass at 575°C. Since τ will increase faster than the interdiffusion coeficient will decrease as the temperature is lowered, treating soda–lime–silica glass in molten KNO_3 at, say, 350°C should result in strengthening with better abrasion resistance. However, as shown by Ward *et al.* (1965) the period of treatment required would be too lengthy to be of practical interest.

Rauscher (1967) has shown that a process analogous to stress relaxation can be produced at room temperature by neutron irradiation of a glass strengthened by $Na^+–Li^+$ exchange. It had been expected that from the *n*

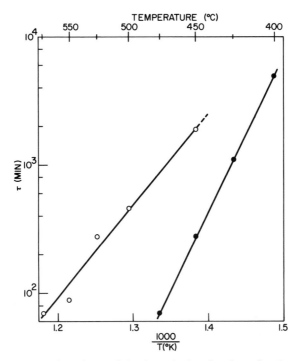

FIG. 13. Temperature dependence of the thermal relaxation time of exchanged 1.1 Na$_2$O–Al$_2$O$_3$–4 SiO$_2$ glass, ○; Corning Code 0080 glass, Na$_2$O–CaO–SiO$_2$, ●.

+ Li^7 = H^3 + He^4 reaction, a differential contraction would occur because of lithium depletion. Density measurements of ion-exchanged samples before and after irradiation did indeed show the expected contraction. The stress profile after irradiation, however, showed a pronounced decrease in stress. Radiation-induced stress release can best be explained by chemical bond rupture caused by recoil atoms. Particularly susceptible would be the regions of stress caused by the large ion-for-small ion exchange, which preceded the irradiation.

C. THERMAL FATIGUE

Because of the need for materials able to withstand sustained use at elevated temperatures, the thermal decay of strength of ion-exchanged glass is of importance. Heating an ion-exchanged glass in the absence of a source of ions results in a loss of strength (Garfinkel, 1967; Garfinkel and King, 1970). Figure 14 shows the effect of air heating on a Li$_2$O–Al$_2$O$_3$–SiO$_2$ glass initially ion exchanged in NaNO$_3$ at 400°C for 4 hr. The thermal decay at temperatures above 400°C is obvious.

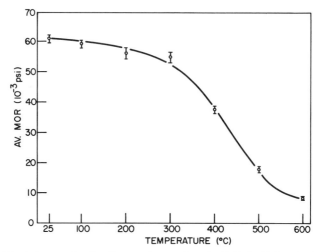

FIG. 14. Abraded strength of sodium exchange 11 mole % Li_2O–16.5 mole % Al_2O_3–72.5 mole % SiO_2 glass after reheating in air for 4 hr at various temperatures.

Assuming that a given profile in the same glass will yield the same modulus of rupture so long as the stress relaxation due to viscous flow is not important, we can calculate the effect of other reheating schedules on samples of $11Li_2O$–$16.5Al_2O_3$–$72.5SiO_2$ treated initially at 400°C for 4 hr. If the final distribution of ions is the same in two samples of the same glass, then either they were reheated in air under identical conditions or they were treated such that the product of the time and interdiffusion coefficient of reheating the first sample was equal to that of the second. That is,

$$D_{11}t_1 = D_{12}t_2, \tag{41}$$

where t is the time of reheat, and D_{11} the diffusion coefficient for the reheating step at temperature T_1.

It follows from Eq. (41) and the temperature dependence of the interdiffusion coefficient that

$$\log(t_1/t_2) = -(E\ddagger/4.6)\,[(1/T_2) - (1/T_1)], \tag{42}$$

where $E\ddagger$ is the measured temperature coefficient of the interdiffusion coefficient.

Figure 15 shows the effect of reheating ion-exchanged samples of the lithia glass at 275°C for times up to 500 hr; a smooth curve was drawn through the points. Both abraded and unabraded results are shown for comparison. The points shown as crosses were calculated with Eq. (42) from the experimental results shown in Fig. 14. All samples were given the same

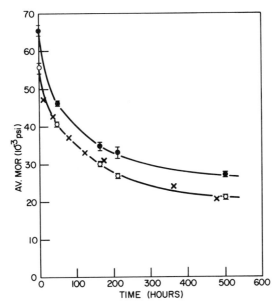

FIG. 15. Strength of ion-exchanged 11 mole % Li$_2$O, 16.5 mole % Al$_2$O$_3$, 72.5 mole % SiO$_2$ after reheating at 275°C for times up to 500 hr. ●, Unabraded; ○, abraded; ×, calculated.

initial ion-exchange treatment. The calculations were made by comparing points at the same fractional decrease in strength. This good agreement indicates that the curve in Fig. 14 adequately describes the thermal decay of strength for all conditions of time and temperature for which viscous flow is unimportant.

Because of the difference in the activation energies for diffusion, K$^+$–Na$^+$ exchanged glasses are about 10 times more stable at 350°C and about 20 times more stable at 300°C than Na$^+$–Li$^+$ exchanged glasses. Potassium-exchanged lithia–alumina–silica glasses have been heated in air for 600 hr at 400°C with no loss in strength. Since diffusion coefficients for the Cs$^+$–Li$^+$ or Rb$^+$–Li$^+$ exchange are smaller by a factor of at least 10^4 than in the Na$^+$–Li$^+$ exchange at 400°C, cesium or rubidium-exchanged lithia glasses should be quite resistant to thermal fatigue. Furthermore, bending these exchanged glasses for long periods of time at high temperatures should result in permanent deformation of the article.

When these ion-exchanged glasses are heated in air at temperatures above the transformation range and cooled, the glass can be weaker than the original annealed sample. This occurs because the surface is now in tension due to the higher expansion glass on the surface.

D. Salt Bath Contamination

In addition to the effect of time and temperature of ion exchange on chemical strengthening (kinetic effects), the effect of the age of the bath (thermodynamic) has been found to have influence on the strength of the treated glass (Varner and Lang-Egelkraut, 1977). As the buildup of the ion in the glass (counter ion) proceeds in the salt bath, they showed that the strength of the glass article decreases. Contamination of the molten salt bath by the counterion reduces the surface concentration at the glass–molten-salt interface of the ion that is being exchanged into the glass. The extent of reduction in strength depends very much on the equilibrium described in Eq. (3).

To overcome this problem, it has been suggested (Garfinkel, 1962) that a "cascade" treatment be used. The ion-exchange reaction is effected stepwise in molten salt baths with decreasing impurities of counter ion. This process can be considered as a multiple diffusion problem in which a semi-infinite solid of zero initial concentration is exposed to various surface concentrations, C_1, C_2, C_3, . . . , C_n, for time intervals 0 to t_1, 0 to t_2, 0 to t_3, . . . , 0 to t_r, respectively. Associated with each time interval $0 < t < t_1$ is an interdiffusion coefficient $D = D_1$. The solution for the first diffusion is the well-known expression

$$c_1(x, t) = c_1 \text{ erfc } [x/2(D_1 t)^{1/2}] \tag{43}$$

where C_1 is the surface concentration and D_1 the interdiffusion coefficient. If the diffusion is terminated at $t = t_1$, then

$$C_1(x, t) = C_2(x, 0) \tag{44}$$

becomes the initial condition for the second diffusion. Then a solution is required of the diffusion equation

$$\partial C_2/\partial t = D_2(\partial^2 C_2/\partial x^2) \tag{45}$$

for the initial condition and the boundary condition

$$C_2(0, t) = C_2, \qquad t > 0. \tag{46}$$

This problem can be solved by means of Green's functions; the solution is expressed as the sum of two integrals. Mathematical manipulation leads to the result

$$C_2(x, t) = C_1 \text{ erfc } \left[\frac{x}{2(D_1 t_1 + D_2 t)^{1/2}} \right] + (C_2 - C_1) \text{ erfc } \left[\frac{x}{2(D_2 t)^{1/2}} \right], \tag{47}$$

which is now the initial condition for the third diffusion; if the diffusion is now terminated at $t = t_2$, then

$$C_2(x, t) = C_3(x, 0) \tag{48}$$

is the initial condition for the third diffusion. Again a solution of the diffusion equation

$$\partial C_3/\partial t = D_3(\partial^2 C_3/\partial x^2) \tag{49}$$

is sought.

By induction the following general expression for n-successive ion-exchange reactions over the time interval $0 < t < t_1$ is obtained:

$$C_n(x_1, t) = \sum_{i=1}^{n} (C_i - C_{i-1})$$
$$\times \operatorname{erfc}\left[\frac{x}{2} \left(\sum_{j=1}^{n-1} D_j t_j - \sum_{j=0}^{i-1} D_j t_j + D_n t \right)^{1/2} \right], \tag{50}$$

where n is the number of successive ion-exchange reactions under consideration and

$$D_0 t_0 = 0, \quad C_0 = 0. \tag{51}$$

The total uptake after n-successive diffusions is given by

$$Q_n = -\int_0^t D_n \left(\frac{\partial C_n}{\partial x} \right)_{x=0} dt = -\int_0^\infty x \frac{\partial C_n}{\partial x} dx \tag{52}$$

or

$$Q_n = \frac{2}{\sqrt{\pi}} \sum_{i=1}^{n} (C_i - C_{i-1}) \left(\sum_{j=1}^{n-1} D_j t_j - \sum_{j=0}^{j-1} D_j t_j + D_n t \right]^{1/2}. \tag{53}$$

Note that for $n = 1$, Eq. (53) reduces to the usual square-root-of-time law for single diffusion.

To illustrate the effect of treating by a "cascade" process, a hypothetical ion-exchange reaction is assumed in which the total time is 2 hr, with a diffusivity equal to 10^{-8} cm²/sec. Molten salt mixtures of $ANO_3 + BNO_3$ are chosen such that the surface concentration C_1, and hence the equilibrium uptake, vary from 7.0 to 10.0 in arbitrary units of concentration for the solvent B.

Concentration–distance profiles calculated from Eq. (50) are shown in Fig. 16. Curves I and IV show the difference between single treatments ($n = 1$) with $C_1 = 10.0$ and 7.0, respectively. Since the diffusivity was assumed constant, the depth is unaffected by diluting the exchangeable

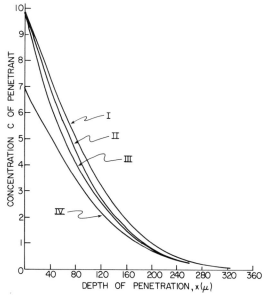

FIG. 16. Comparison of several cascade schemes with single exchange processes calculated from Eq. (50) with $D_{12} = 10^{-8}$ cm^{-2} sec^{-1} at temperature T. Curve I: $C_i = 10.0$, $t = 2.0$ hr; II: $C_i = 7.0, 8.0, 9.0, 10.0$, $t_1 = t_2 = t_3 = t_4 = 30$ min; III: $C_i = 7.0, 8.5, 10.0$, $t_1 = 40$, $t_2 = 75$, $t_3 = 5$ min; IV: $C_i = 7.0$, $t = 2.0$ hr.

BNO$_3$ with the salt ANO$_3$. It is apparent that the surface concentration and the total amount exchanged (the area under the curve) is less in case IV than in case I. In schemes II ($n = 4$) and III ($n = 3$), not only is the surface concentration the case as in I, but the total amount exchanged approaches that in I. It is interesting to note that in curve III, the effect of the last 5 min treatment is mainly to bring the surface concentration up to the prescribed ten arbitrary units.

VII. Glass Ceramics

Polycrystalline glass ceramics can also be strengthened by the techniques described for glass. This is possible because they are free from open flaws and are nonporous. The ion-exchange properties of glass ceramics are not too dissimilar from those of the parent glass as shown by Bartholomew and Garfinkel (1977) for Corning Code 9608 glass ceramic—a β-spodumene structure. Diffusion rates (interdiffusion) are about a factor of 3 slower in the glass ceramic in comparison with the glass. Glass ceramics also can be strengthened both by expansion difference and by ion stuffing. Both techniques can be subdivided further into those materials that form solid solutions on replacement of one ion by another, or where phase transformation occurs. The wide variety of metastable and stable phases that can be crystallized from a glass and the considerable compositional variations possible due to solid solution form the bases for these strengthening techniques. Because of the nature of these materials, normally higher temperatures are needed for ion exchange in a glass ceramic in comparison with the glass precursor.

Karstetter and Voss (1967) reported that a fine-grained glass ceramic containing a large proportion of β-spodumene solid-solution crystals was strengthened by immersion in both molten sodium- and potassium-containing salt baths. The resulting sodium or potassium ion replacement of lithium ions produced a compressive surface layer. Strengths in excess of 100,000 psi were measured on abraded samples (modulus of rupture). Similarly, stuffed β-quartz solid-solution glass ceramics derived from the crystallization of Li_2O–Al_2O_3–SiO_2 glasses were also strengthened by K^+–Li^+ exchange. However, in metastable β-quartz material no strengthening was seen when Na^+–Li^+ exchange was tried; this is in contrast to increased strength seen in stable β-quartz solid-solution glass ceramics.

Beall et al. (1967), as well as Duke et al. (1966), showed still other and more complex ways in which ion exchange can be used to generate surface compression, some of which involve a 2 Li^+–Mg^+ exchange in which the highest strengths obtained by solid solution "stuffing" were reported. Table I shows the compositions, and Table II the strengths obtained (all samples were tumble abraded for 15 min with 30 grit SiC) by different exchange baths. The last three compositions represent strengthening by stuffing as a result of a phase transformation induced by ion exchange. The kalsilite ($KAlSiO_2$) formed at the surface has a higher expansion coefficient than the nepheline body that would normally result in surface tensile forces and reduced strength. However, the stuffing process produces a net surface compression to yield strengths as high as 200,000 psi.

Ernsberger (1975) has proposed a novel method for strengthening glass ceramics by which surface compression is developed by first nucleating a surface layer, then increasing the temperature so that crystallization of the surface is complete, and the interior is nucleated. On a further increase in temperature, the interior crystallizes. However, the already crystalline surface is rigid and so does not flow to relieve the compressive stress generated by shrinkage of the interior. A compressive surface stress that occurs during the second stage of heat treatment persists essentially unaltered as the article is cooled to room temperature. This gives an article in which the final composition and physical state of the glass ceramic are uniform throughout. There are no concentration gradients or expansion differences; hence, the surface compression should be unaffected by temperature changes. A simple and effective way to achieve control of the timing was

TABLE I

COMPOSITIONS OF REPRESENTATIVE GLASS CERAMICS

Composition	1	2	3	4	5	6	7	8	9	10
SiO_2	56	51	60	69	50	71	40	44	43	41
Al_2O_3	20	26	—	21	29	18	31	32	32	32
Li_2O	—	—	—	5	—	2	—	—	—	11
Na_2O	—	—	5	—	—	—	4	17	14	9
K_2O	—	—	—	—	—	—	18	—	4	—
MgO	15	5	—	—	12	5	—	—	—	—
ZnO	—	8	—	—	—	—	—	—	—	—
BaO	—	—	35	—	—	—	—	—	—	—
TiO_2	9	3	—	5	9	—	7	7	7	7
ZrO_2	—	7	—	—	—	4	—	—	—	—
Crystal phases	α-Quartz spinel enstatite	α-Cristobalite spinel ZrO_2	Barium silicate	β-Spodumene rutile	β-Quartz $MgTi_2O_5$	β-Quartz ZrO_2	Synthetic Kaliophilite	Nepheline anatase	Nepheline anatase	Nephelin anatase
Expansion coefficient (23–300°C)	100	166	195	15	40	16	140	115	120	125

TABLE II

STRENGTH OF ION-EXCHANGED GLASS CERAMICS (ABRADED)

Composition	Bath	Temperature (°C)	Time (Hr.)	Exchange	Surface phase	Modulus of rupture (10^3 psi)
A) Glass ceramics strengthened by expansion difference						
1	Li_2SO_4	950	24	$2 Li^+ \to Mg^{2+}$	β-Quartz ss	143
2	Li_2SO_4	950	24	$2 Li^+ \to Mg^{2+}$	β-Quartz ss	spalling
3	52% KCl	850	2	$2 K^+ \to Ba^{2+}$	Glass	40
	48% K_2SO_4					
5	Li_2SO_4	850	8	$2 Li^+ \to Mg^{2+}$	β-Quartz ss	160
B) Glass ceramics strengthened by stuffing mechanism						
4	85% $NaNO_3$	580	16	$Na \to Li^+$	β-Spodumene ss	90
	15% Na_2SO_4					
5	Li_2SO_4	850	8	$2 Li \to Mg^{2+}$	β-Quartz ss	160
6	52% KCl	750	8	$K^+ \to Li$	β-Quartz ss	46
	48% K_2SO_4					
7	52% KCl	730	8	$K \to Na^+$	Kaliophilite	100
	48% K_2SO_4					
8	52% KCl	730	8	$K \to Na^+$	Kalsilite	90
	48% K_2SO_4					
9	52% KCl	730	8	$K^+ \to Na^+$	Kalsilite	179
	48% K_2SO_4					
10	52% KCl	730	8	$K \to Na^+$	Kalsilite	200
	48% K_2SO_4					

to carry out the entire heat treatment in an atmosphere of steam. The small amount of water that entered the glass has a catalytic effect on nucleation and crystallization. A lithium aluminosilicate glass was strengthened by this technique to a strength of approximately 40,000 psi over the temperature range from room temperature to 700°C.

VIII. Multistep Strengthening Techniques

A. TWO-AND THREE-STEP METHODS

In the past decade the patent literature has contained many methods of strengthening glass that are actually combinations of methods. These can be either combinations of ion exchange and thermal toughening or of ion exchange and acid-etching, or the most interesting multiple ion-exchange techniques. The objective of these techniques is to combine the good properties of each method so as to generate a high degree of strengthening with favorable fracture behavior, better thermal stability, durability, and deeper compression layer to be able to overcome the surface flaws and hence reduce the delayed breakage. Many possible combinations have been investigated and shown to result in increased abraded strength. Table IIIa

TABLE III
Multistep Techniques for Increased Abraded Strength

Combination	Glass	Strain pt. (°C)	(a) Two-step methods Treatment steps		Abraded strength (psi)	Reference
			First	Second		
1. Thermal tempering	(a) Na_2O–CaO–SiO_2	Not given	Air quenched	KNO_3:400°C, 24 hr	43,167	Hess et al. (1966)
2. Ion exchange below the strain point	(b) Na_2O–Li_2O–Al_2O_3–SiO_2	506	Quenched in molten salt bath	$NaNO_3$:450°C, 2 hr	Reduced delayed breakage of tumblers	Megles (1969)
1. Small ion for large ion above the strain point 2. Large ion for small ion below the strain point	Na_2O–Al_2O_3–P_2O_5–SiO_2	580	75 mole% Li_2SO_4– 25 mole% Na_2SO_4:750°C, 10 min	85 mole% $NaNO_3$– 15 mole% Na_2SO_4:450°C, 2 hr	67,000	Eppler and Garfinkel (1970)
1. Small ion for large ion above the strain point 2. Reheat below the strain point	Na_2O–Al_2O_3–P_2O_5–SiO_2	576	75 mole% Li_2SO_4– 25 mole% Na_2SO_4:750°C, 10 min	Air reheat:425°C, 4 hr	47,000	Olcott (1968)
1. Small ion for large ion above the strain point 2. Large ion for large ion below the strain point	Na_2O–Al_2O_3–SiO_2	553	80 mole% Li_2SO_4– 20 mole% LiCl:750°C, 5 min	$LiNO_3$:350°C, 1 hr	37,000	Chisholm et al. (1966)
1. Large ion for small ion below the strain point 2. Large ion for small ion below the strain point	Na_2O–Li_2O–Al_2O_3–SiO_2	566	85 mole% $NaNO_3$– 15 mole% Na_2SO_4:450°C, 2 hr	KNO_3:450°C, 1 hr	½ lb. steel ball broke plate at 174 in. after 2nd step, 78 in. after 1st step	Marusak (1968)

Combination	Glass	Strain pt. (°C)	First	Second	Third	Abraded strength (psi)	Reference
1. Large ion for small ion above the strain point 2. Large ion for small below the strain point	Na_2O–K_2O–Al_2O_3–SiO_2	581	52 mole% KCl–48 mole% KNO_3:525°C, 4 hr K₂SO₄:750°C, 20 min			18,000	Grego and Howell (1973)

(b) Three-step methods — Treatment steps

Combination	Glass	Strain pt. (°C)	First	Second	Third	Abraded strength (psi)	Reference
1. Small ion for large ion above strain point 2. Small ion for large ion below the strain point 3. Large ion for small ion below the strain point	Na_2O–Al_2O_3–P_2O_5–SiO_2	553	80 mole% Li_2SO_4–20 mole% LiCl:760°C, 15 min	$LiNO_3$:360°C, 1 hr	$NaNO_3$:400°C, 4hr	57,000	Garfinkel and Olcott (1971)
1. Large ion for small ion above the strain point 2. Air reheat above the strain point 3. Large ion for small ion below the strain point	Na_2O–K_2O–Al_2O_3–SiO_2	581	$K_2Cr_2O_7$:625°C, 1 hr	Air reheat: 700°C, 20 min	KNO_3: 525°C, 8hr	43,800	Grego and Howell (1973)
1. Small ion for large ion above the strain point 2. Air reheat below the strain point 3. Large ion for small ion below the strain point	Na_2O–Al_2O_3–P_2O_5–SiO_2	553	80 mole% Li_2SO_4–20 mole% LiCl: 760°C, 5 min	Air reheat: 400°C, 4 hr	$NaNO_3$:400°C, 4 hr	71,400	Garfinkel (1966) (see fig. 20)

257

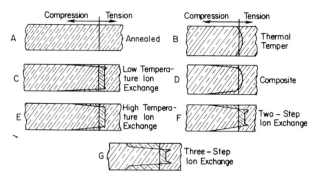

FIG. 17. Schematic stress profiles produced as a result of different treatments: (a) annealed, (b) thermal temper, (c) low-temperature ion exchange, (d) composite-thermal plus chemical temper, (e) high-temperature ion exchange, (f) two-step ion exchange, (g) three-step ion exchange.

shows some of the two-step methods. Table IIIb illustrates some three-step methods. Some typical stress profiles obtained under such treatments are shown schematically in Fig. 17. The more steps involved, the more complex the stress profile tending to a W shape.

The effect of reheating below the strain point when a small ion is exchanged for a large ion above the strain point results in smoothing out the profiles, such that the concentration of the large ion increases at the surface. This is equivalent to a stuffing exchange. Figure 18 (Olcott, 1968) illustrates the effect of time of reheat and temperature on a glass of 50% SiO_2, 20% Al_2O_3, 10% P_2O_5, 18% Na_2O, 1% K_2O, and 1% Li_2O (wt %) which had been exchanged in a 75 mole % Li_2SO_4–25 mole % Na_2SO_4 bath at 750°C for 10 min as a first step prior to the reheating at 450 and 525°C. A

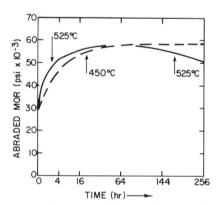

FIG. 18. Effect of reheat on 50% SiO_2, 20% Al_2O_3, 10% P_2O_5, 18% N_2O, 1% K_2O, and 1% Li_2O (wt %) glass after exchange in Li_2SO_4–$NaSO_4$ bath at 750°C, 10 min. Reheated at 450°C and 525°C.

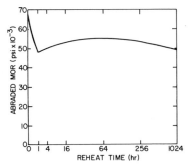

FIG. 19. 50% SiO$_2$, 23% Al$_2$O$_3$, 7% P$_2$O$_5$, 18% Na$_2$O, 1% K$_2$O, and 1% Li$_2$O (wt %) glass treated in a Li$_2$SO$_4$–Na$_2$SO$_4$ bath 750°C, 10 min followed by 2 hr at 450°C in NaNO$_3$–Na$_2$SO$_4$ bath. Effect of reheating at 500°C.

glass similar in composition except for 23% Al$_2$O$_3$, 7% P$_2$O$_5$, was treated using the same first step (Eppler and Garfinkel, 1970). However, this was then followed by a second step in molten 85% NaNO$_3$–15% Na$_2$SO$_4$ for 2 hr at 450°C. Reheating this sample at 500°C resulted in a variation of MOR with treatment time as shown in Fig. 19.

An actual stress profile of a sample treated by the third combination described in Table IIIb for three-step methods (Garfinkel, 1966) is shown in Fig. 20. This profile is very similar to the schematic three-step profile shown in Fig. 17. The depth of the compression layer is approximately 0.014 mils. Using just the first two steps on the same glass composition results in a strength of about 40,000 psi.

It is obvious from Table III that many other combinations are possible. What has to be taken into account is not only the maximum modulus of rupture required, but also the type of breakage (violent or nonviolent) as well as the need to withstand elevated temperatures and maintain strength.

IX. Glass Composition—Effect on Strength

Important as the effect of time and temperature is on the abraded strength of a sample, that of the glass composition can be far greater. This is illustrated in Table IV. In some cases the glass shows reasonably high unabraded, but poor abraded strength. Obviously, not every glass can be strengthened by ion exchange.

A. SODA–SILICA

It is obvious that there is some level of alkali concentration required to obtain sufficient ion exchange for strengthening. But the effect of multivalent oxides in the role of network formers appears to play the largest role. Ion exchange of K$^+$ for Na$^+$ in simple Na$_2$O–SiO$_2$ has been found to lead

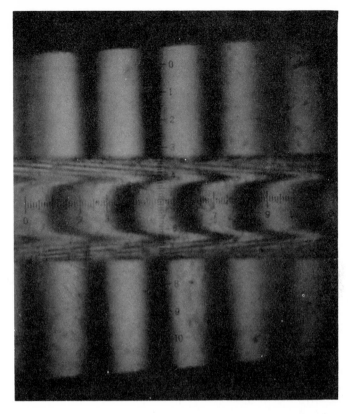

FIG. 20. Stress profile after three-step treatment: (a) Li_2SO_4–LiCl at 762°C for 5 min; (b) air reheat 400°C for 4 hr; (c) $NaNO_3$ at 400°C for 4 hr; 60% SiO_2, 15% Na_2O, 1%, K_2O, 5% P_2O_5, 18% Al_2O_3 glass (wt %).

to little or no compressive stress at 375°C and below (Hahnert and Hinz, 1967; Spoor and Burggraaf, 1966). Stress relaxation of the surface is the major cause.

B. SODA–LIME–SILICA

There has been a great deal of effort spent in trying to strengthen soda–lime–silica glasses because of the commercial advantages of such a glass system. Ward *et al.* (1965) reported that although it is possible to increase the strength of a soda–lime–silica glass by K^+ for Na^+ ion exchange, the period of time required to get a sufficiently deep layer is too long to be commercially practical. In addition the stress relaxation in soda–lime

TABLE IV

STRENGTHENING BY POTASSIUM ION EXCHANGE[a]

Composition (wt %)	Production glasses				Experimental Glasses	
	Bulb	Sheet	Borosil	Lead	Borosil	Alum–Sil
SiO_2	73	72.5	81	56	75	54.5
Na_2O	16.5	15	4	4	15	16
K_2O	—	—	—	—	—	2
Al_2O_3	1	1	2	—	—	19
B_2O_3	—	—	12	10	10	22
CaO	5	9	—	—	—	2
MgO	3.5	2.5	—	—	—	—
PbO	—	—	—	30	—	—
TiO_2	—	—	—	—	—	4.5
K^+ Exchanged (mg/cm^2)	0.5	0.15	0.04	0.2	0.16	0.5
MOR (kg/mm^2) No abrasion	49 ± 20	35 ± 15	20 ± 6	8 ± 2	24 ± 9	61 ± 1
150 Grit	12 ± 4	6 ± 2	7 ± 1	4 ± 1	11 ± 6	48 ± 4

[a]Treated for 16 hr at 400°C.

glasses has been discussed elsewhere in this review. Nordberg *et al.* (1964) have shown the comparison of additions of CaO to simple soda–lime–silica glasses with additions of Al_2O_3 to soda–alumina–silica glasses. It should be noted that the amount of exchange decreases with increasing CaO addition, but little, if any, with increasing Al_2O_3 content. This finding emphasizes the fact that the depth of exchange layer is small. Attempts to increase the depth of the layer by multiple treatment have been made (see Table IIIa) in the soda–lime–silica glasses, but very high abraded strengths have not been obtained.

C. SODA–ALUMINA–SILICA AND SODA–ZIRCONIA–SILICA

Nordberg *et al.* (1964) also showed abraded MOR increased as a function of Al_2O_3 in 20% and 10% Na_2O-containing Na_2O–Al_2O_3–SiO_2 glasses when treated for 16 hr at 380°C in KNO_3. Lowering the Na_2O in the glass reduces the abraded strength slightly, but did not change the direct correlation between strength and alumina content. In fact not only were they able to show the increased abraded strength for the K^+–Na^+ exchange, but a similar effect was seen for the Na^+–Li^+ exchange in lithia–alumina–silica glasses. A linear correlation was found between abraded strength and alumina content for both potassium exchanged soda–alumina–silica glasses

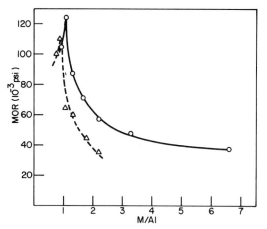

FIG. 21. Dependence of the strength of ion-exchanged Al_2O_3 glasses on the M(Li or Na)/Al ratio. ○, Na_2O–Al_2O_3–SiO_3; △, Li_2O–Al_2O_3–SiO_2.

and sodium exchanged lithia–alumina–silica glasses. In both cases the strength was a maximum at $M/Al = 1.0$ (M = Na or Li) (see Fig. 21). Burggraaf and Cornelissen (1964) also reported a maximum in the strength of alumina containing glass at the M/Al ratio equal to one. At this ratio the glasses should contain no nonbridging oxygen ions. Alkali aluminosilicate glasses show a maximum in electrical conductivity (Isard, 1959; Tsekhom-skii *et al.*, 1963; Terai, 1969), self-diffusion coefficient of alkali ion (Terai, 1969), and interdiffusion properties at this molar ratio of unity (Burggraaf and Cornelissen, 1964). It has been postulated (Varshneya and Milburg, 1974) that glasses with fewer nonbridging oxygens have a more covalent structure than glasses that have an appreciable number of nonbridging oxygens. This can be interpreted on the basis that glasses without nonbridg-ing oxygens may withstand higher stresses without undergoing plastic or viscoelastic deformation.

A similar correlation between strength and zirconia content in soda–zir-conia–silica glasses was reported by Nordberg *et al.* (1964). These data are shown in Fig. 22. Again MOR data show retention of strength after abra-sion. The zirconia-containing glasses are of less commercial interest than the corresponding alumina glasses because of melting difficulties.

Although lithium glasses cost more than soda glasses, the rate of exchange of Na^+–Li^+ is about 10 times greater than the K^+–Na^+ exchange. Therefore, in order to obtain sufficient depth of exchange to withstand severe abrasion during use, a soda–alumina–silica glass must be treated in molten KNO_3 at temperatures 100°–150°C higher than those for the corre-sponding lithia–alumina–silica glass treated in molten $NaNO_3$.

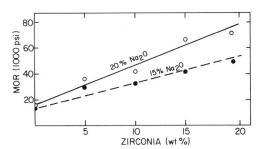

FIG. 22. Dependence of the strength of ion-exchanged $Na_2O-ZrO_2-SiO_2$ glasses on the ZrO_2 content of the glass. Treatment: 16 hr at 380°C in KNO_3, 20% Na_2O; 15% Na_2O.

D. SODA–BOROSILICATE

Chemical strengthening has been studied using $Na_2O-B_2O_3-SiO_2$ glasses (Brungs and McCartney, 1975) with a constant 15 mole % Na_2O with the Na-to-B ratio varied from 1.42 to 0.49. Although their compositions were close to the composition region in which phase separation occurs, Brungs and McCartney found no phase separation in these glasses. The authors selected silver for sodium ion exchange below the strain point, mainly because there is much less "burnout" of the silver ions during electron microprobe analysis (the technique they used to check ion exchange) than the more conventional potassium. They used a 20 mole % $AgNO_3$ and 80 mole % KNO_3 molten salt bath; the potassium exchange rate was negligible due to the great difference in diffusion rates of potassium and silver and the high selectivity of the glass for silver relative to potassium. The samples were abraded using 27 μm alumina powder. Their data showed that at a Na : B ratio of 1.42 the treated samples were four times stronger (approximately 60,000 psi) than the untreated, while at a Na : B ratio of 0.49 this strength increase was only about twofold. The modulus of rupture was proportional to the silver penetration up to approximately 33 μm depth. After that the strength remained constant.

E. ADDITIONS OF ZnO OR P_2O_5

Other oxide additions that have been claimed to have a beneficial effect on strengthening include ZnO (Rinehart, 1966) and P_2O_5 (Rinehart, 1967). Rinehart has carried out Na^+-Li^+ and K^+-Li^+ exchange in both lithia–zinc oxide–alumina–silica and lithia–phosphorus pentoxide–alumina–silica glasses. The incorporation of P_2O_5 greatly enhanced ion exchange and therefore resulted in a deeper layer. This effect is enhanced by additions of Na_2O to the initial base glass. Replacement of P_2O_5 by MgO in these glasses reduced the depth of the surface compressive stress zone by almost one-half for identical molten salt treatments. The abraded strength of phosphorus pentoxide-containing glasses after exchange at 480°C for 36 min was in the region of 38,000 psi. Similar advantages are claimed for the glasses containing ZnO.

F. MIXED ALKALI GLASSES

It has been observed on several occasions that glasses, which initially contain both small and large alkali ions, develop deeper compressive surface layers upon ion exchange than those that contain only small ions. A detailed study of this effect has been reported by Ohta (1975) based on a series of glass compositions based on xNa$_2$O, $(1 - x)$K$_2$O, MgO, SiO$_2$, where MgO was 16 mole % and SiO$_2$ 68 mole %. Exchange reactions were carried out using a KNO$_3$ melt for 20 hr at 420, 440, 460, or 490°C. Stress profiles were determined optically using a Berek compensator. The stress profiles showed that the depth of the compressive layer increased as the potassium content in the original glass increased. However, although the depth of the compressive layer increased with K$_2$O content, the maximum compressive stress decreased. The amount of potassium that had diffused into the glass decreased with increase in potash content. Ohta found that with increased potassium content in the parent glass the interdiffusion coefficient increased while the temperature coefficient of exchange decreased. Glasses of high Na$_2$O content showed stress relaxation even at 420°C as indicated by the stress profiles, which revealed maximum stress in the interior of the sample. The author postulates that in a glass containing both soda and potash, exchange of potassium ion for sodium ion takes place inside the glass at the same time as at the surface. The presence of the compressive stress after ion exchange is put forth as evidence that the probability of a potassium ion in a glass containing both sodium and potassium ions jumping from a potassium site to a neighboring sodium site (produces compression) is greater than is for a sodium jumping to a potassium site (tensile stress).

G. SILVER AND COPPER ION EXCHANGE

Except for the work of Brungs and McCartney (1975), the strengthening of glass by the ion-exchange technique has been concentrated on the alkali metal ions Li$^+$, Na$^+$, and K$^+$. Several commercially available glass compositions were treated by Garfinkel (1962) in AgNO$_3$–AgCl bath at 325°C for 4 hr. The MOR of cane samples, given a 150 grit abrasion, are shown in Table V. Except for the Li$_2$O–Al$_2$O$_3$–SiO$_2$ composition, the glasses were slightly yellow in color after the exchange. After 16 hr in molten KNO$_3$ at 380°C, Corning Code 0080 and Code 0088 gave 17,500 and 26,500 psi values, respectively, for MOR (150 grit abrasive). The low-alkali content in Code 7740 is responsible for the low MOR value in that glass.

TABLE V

THE MOR OF CANE SAMPLES[a]

Glass	Modulus of rupture (psi) (−150 grit abraded)
Corning Code 0080	40,600 ± 3,700
Corning Code 0088	59,200 ± 2,300
Corning Code 0281	23,900 ± 6,100
Corning Code 7740	11,000 ± 900
11.4 mole % Li_2O, 16.8 mole % Al_2O_3, 71.6 mole % SiO_2	86,700 ± 4,400

[a]Samples were treated for 4 hr. at 325°C in 75 mole % $AgNO_3$, 25 mole % AgCl bath.

In comparison to strength data obtained with silver exchange, the results obtained for glasses treated in a cuprous chloride melt were much lower. Because the Cu(I) ion has a comparable radius to the Na^+ ion, the strengthening effect should be observed for glasses in which Cu(I) for Li^+ ion exchange occurs. The Li_2O–Al_2O_3–SiO_2 glass was treated for 4 hr at 480°C in cuprous chloride; the resulting strength (abraded) was 21,500 ± 300 psi. A slight increase in strength, but much less than seen with silver or with Na^+ ion exchange at 400°C (~70,000 psi).

Although Rb^+−Li^+ and Cs^+−Li^+ exchange kinetics have been reported (Garfinkel, 1972), no strength measurements were made. The long times and high temperatures required for appreciable amounts of exchange to occur meant that depths of exchange were small while stress relaxation was considerable. Frischat *et al.* (1974) measured diffusion properties for Rb^+−Na^+ exchange; again no strength or stress data were reported.

X. Applications

The initial incentive for commercial application of chemically strengthened glass was in automobile windshields. In the middle 1960s American Motors produced a car with a laminated, chemically strengthened windshield, while Ford Motor Company used chemically strengthened glass as a backlight in some convertibles. However, improvements in methods of manufacturing annealed-laminated windshields coupled with the higher cost of chemically strengthened glass led to the disappearance of that business opportunity. Perhaps present lower weight requirements of today will rekindle interest.

The Federal Drug Administration (Federal Register, 1971) gave great impetus to chemical strengthening of ophthalmic glass by publishing regulations defining a minimum impact condition for all glasses sold in the U.S. Today most glass prescription eyewear that is sold is given a 16-hr (overnight) treatment in KNO_3 at about 400°C. Because the exchange occurs below the strain point of the glass, there is no change in surface figure of the prescription lens. Lenses are tested by dropping a steel ball from a height of 50 in. A comparison of ball-impact test results [chemically strengthed, tempered, and plastic (C.R. 39)] is shown in Fig. 23 as a function of lens power (Chase *et al.*, 1973). The lower the lens power, the lower the lens center thickness.

Chemically strengthened glass has several advantages over thermally tempered glass. Much thinner glass can be strengthened by ion exchange; there is also no distortion of the sample during the strengthening process. In addition, strengths as much as fivefold greater can be achieved by ion exchange in comparison with samples made by thermal tempering. Unusual shapes are more readily strengthened by ion exchange. However, once an article is strengthened by ion exchange, it can no longer be subjected to any of the traditional finishing processes such as flameworking, cutting, or grinding. Because of the diffusion characteristics of the ions involved, most chemically strengthened materials have high use temperatures (>200°C). In addition, because of the present state of knowledge, the correct treatment can be chosen such that when the chemically strengthened article does fracture it can do so in a controlled manner. That is, the fracture can be violent, breaking into many tiny particles, or it can be mild, resulting in just two or three large pieces.

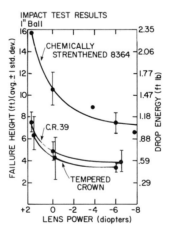

FIG. 23. Impact strength of chemically strengthened Corning Code 8364 glass, a tempered crown glass and CR39 (polycarbonate plastic lens) using 1 in. stainless-steel ball.

One of the major disadvantages of ion exchange to strengthen glass is the cost involved, both in time and materials. As mentioned not all glasses can be chemically strengthened; in fact many of the better glasses contain Li_2O and Al_2O_3 and are difficult or expensive to melt. There is no doubt that chemical strengthening of glass has added a new dimension to the usefulness of one of man's oldest materials. Such applications include windows for both airplane cockpits and space vehicles, laboratory pipettes, architectural panels, ophthalmic glasses, consumer ware, and glass computer tape reels, to name just a few.

References

Abou-el-Leil, M., and Cooper, A. R. (1978). *J. Am. Ceram. Soc.* **61**, 131.

Acloque, P. and Tachon, J. (1962). *Symp. Mech. Strength Glass and Ways of Improving It* pp. 667–704. Union Scientifique Continentale du Verre, Charleroi, Belgium.

Adams, R., and McMillan, P. W. (1977). *J. Mater. Sci.* **12**, 643.

American Society for Testing Materials (1975). "Flexure Testing of Glass," Annual Book of ASTM Standards, Part 17, pp. 104–113. Philadelphia, Pennsylvania.

Bartholomew, R. F. (1966). *J. Phys. Chem.* **70**, 3442.

Bartholomew, R. F., and Garfinkel, H. M. (1977). *In* "Non-Crystalline Solids" (G. H. Frischat, ed.), pp. 459–464. Trans Techn. Publ., Aedermannsdorf, Switzerland.

Beall, G. H., Karstetter, B. R., and Rittler, H. L. (1967). *J. Am. Ceram. Soc.* **50**, 181.

Berger, E. (1924). *Naturwissenshaften* **25**, 79.

Borom, M. P., and Hanneman, R. E. (1967). *J. Appl. Phys.* **38**, 2406.

Brungs, M. P., and McCartney, E. R. (1975). *Phys. Chem. Glasses* **16**, 44.

Burggraaf, A. J. (1966). *Philips Res. Rep. Suppl.* No. 3.

Burggraaf, A. J., and Cornelissen, J. (1964). *Phys. Chem. Glasses* **5**, 123.

Charles, R. J. (1958). *J. Appl. Phys.* **29**, 1549.

Chase, G. A., Kozlowski, T. R., and Krause, R. P. (1973). *Am. J. Optomet. Arch. Am. Acad. Optomet.* **50**, 470.

Chisholm, R. S., Sleighter, G. E., and Ernsberger, F. M. (1966). U.S. Patent 3,287,201.

Compaan, K., and Haven, Y. (1956). *Trans. Faraday Soc.* **52**, 786.

Cooper, A. R., and Krohn, D. A. (1969). *J. Am. Ceram. Soc.* **52**, 665.

Cornelissen, J., Piesolinger, G., and deRijk, A. M. M. (1967). *Symp. Surface Glass and Its Mod. Treatments,* Luxemburg. pp. 145–163. Union Scientifique Continentale du Verre, Charleroi, Belgium.

Crank, J. (1956). "Mathematics of Diffusion." Oxford Univ. Press, London and New York.

Doremus, R. H. (1962). *In* "Modern Aspects of the Vitreous State" (J. D. Mackenzie, ed.), Vol. II, pp. 1–71, Butterworths, London.

Doremus, R. H. (1964). *J. Phys. Chem.* **68**, 2212.

Doremus, R. H. (1973). "Glass Science." Wiley, New York.

Duke, D. A., Beall, G. H., MacDowell, J. F., and Karstetter, B. R. (1966). *Int. Congr. Ind. Chem., 36th Brussels* Paper S18.

Eppler, R. A., and Garfinkel, H. M. (1970). U.S. Patent 3,533,888.

Ernsberger, F. M. (1975). *Bull. Am. Ceram. Soc.* **54**, 533.

Eshelby, J. D. (1956). *Solid State Phys.* **3**, 79.

Faile, S. P., and Roy, R. (1971). *J. Am. Ceram. Soc.* **54**, 532.

Fed. Regist. (1971). **36** (95), 8939.

Fletcher, P. C., and Tillman, J. J. (1964). *J. Am. Ceram. Soc.* **47**, 382.

Frischat, G. H. (1970). *J. Non-Crystall. Solids* **3**, 407.

Frischat, G. H. (1975). "Ionic Diffusion in Oxide Glasses." Trans. Tech. Publ., Aedermannsdorf, Switzerland.

Frischat, G. H., Eichhorn, U., Kirchmeyer, R., and Salge, H. (1974). *Glastech. Ber.* **47**, 107.

Garfinkel, H. M. (1962). Corning Glass Works Internal Rep.

Garfinkel, H. M. (1966). Corning Glass Works Internal Rep.

Garfinkel, H. M. (1967). *Symp. Surfaces Glass and Its Mod. Treatments, Luxembourg* pp. 165–180. Union Scientifique Continentale du Verre, Charleroi, Belgium.

Garfinkel, H. M. (1968). *J. Phys. Chem.* **72**, 4175.

Garfinkel, H. M. (1969) *Glass Ind.* **50**, 28, 74.

Garfinkel, H. M. (1970). *Phys. Chem Glasses* **11**, 151.

Garfinkel, H. M. (1972). *In* "Membranes" (G. Eisenman, ed.), Vol. I, pp. 179–247. Dekker, New York.

Garfinkel, H. M., and King, C. B. (1970). *J. Am. Ceram. Soc.* **53**, 686.

Garfinkel, H. M., and Olcott, J. S. (1971). U.S. Patent 3,630,704.

Garfinkel, H. M., and Rauscher, H. E. (1966). *J. Appl. Phys.* **37**, 2169.

Garfinkel, H. M., Rothermel, D. L., and Stookey, S. D. (1962). *Adv. Glass Technol.*, (Tech. Papers of *Int. Congr. Glass, 6th, Washington, D.C.*), pp. 404–411. Plenum Press, New York.

Grego, P., and Howell, R. G. (1973). U.S. Patent 3,751,238.

Griffith, A. A. (1920). *Phil. Trans. R. Soc. London* **A221**, 163.

Grubb, E. F., and LaDue, A. W. (1974). U.S. Patent 3,844, 754.

Guillement, C., Pierre-dit-Mery, J. M., and Bonnetin, A. (1967). *Symp. Surfaces Glass and Its Mod. Treatments, Luxembourg,* pp. 181–204. Union Scientifique Continentale du Verre, Charleroi, Belgium.

Hahnert, M., and Hinz, W. (1967) *Silikattechn.* **18**, 377.

Hale, D. K. (1968). *Nature (London)* **217**, 1115.

Haven, Y., and Stevels, J. M. (1957). *Proc. Int. Congr. Glass 4th* pp. 343–347. Imprimerie Chaix, 20 rue Bergene, Paris.

Haven, Y., and Verkerk, B. (1965). *Phys. Chem. Glasses* **6**, 38.

Hess, A. R., Sleighter, G. E., and Ernsberger, F. M. (1966). U.S. Patent 3,287,200.

Helfferich, F. (1962). *J. Phys. Chem.* **66**, 39.

Helfferich, F., and Plesset, M. S. (1958). *J. Chem. Phys.* **28**, 418.

Hood, H. P., and Stookey, S. D. (1961). U. S. Patent 2,998,675.

Isard, J. O. (1959). *J. Soc. Glass Technol.* **43**, 113.

Ish-Shalom, M., and Winitzer, S. (1972). *Glass Technol.* **13**, 148.

Janz, G. J. (1967). "Molten Salt Handbook." Academic Press, New York.

Kamita, K. (1930). U.S. Patent 1,782,169.

Kane, W. T. (1973). *In* "Microprobe Analysis" (C.A. Anderson, ed.), pp. 241–270. Wiley, New York.

Kane, W. T., and Williams, J. P. (1971). *Proc. Int. Congr. Glass, 9th Versailles* pp. 285–302.

Karreman, G., and Eisenman, G. (1962). *Bull. Math. Biophys.* **24**, 413.

Karstetter, B. R. and Voss, R. O. (1967). *J. Am. Ceram. Soc.* **50**, 133.

Keppeler, G. (1927). *Glastech. Ber.* **5**, 97.

Kielland, J. (1935). *J. Soc. Chem. Ind.* **54**, 232T.

Kistler, S. S. (1962). *J. Am. Ceram. Soc.* **45**, 59.

Koslowski, T. R., and Bartholomew, R. F. (1968). *Inorg. Chem.* **7**, 2247.

LaCourse, W. C. (1972). *In* "Introduction to Glass Science" (L. D. Pye *et al.*, eds.), pp. 451–512. Plenum Press, New York.

Lewek, S. S. (1968). U.S. Patent 3,395,999.
Marusak, F. J. (1968). U.S. Patent 3,410,673.
Meagles, J. E. (1969). U.S. Patent 3,445,316.
Meistring, R., Frischat, G. H. and Hennicke, H. W. (1976). *Glastech. Ber.* **49**, 60.
Miller, W. L., and Gordon A. R. (1931). *J. Phys. Chem.* **35**, 2785.
Mochel, E. L., Nordberg, M. E., and Elmer, T. H. (1966). *J. Am. Ceram. Soc.* **49**, 585.
Moitra, A. K., Gupta, P. K., and Kumar, S. (1971). *Proc. Int. Congr. Glass, 9th, Versailles* pp. 107–117.
Morey, G. W. (1954). "The Properties of Glass," 2nd ed. Van Nostrand-Reinhold, Princeton, New Jersey.
Nordberg, M. E., Mochel, E. L., Garfinkel, H. M., and Olcott, J. S. (1964). *J. Am. Ceram. Soc.* **47**, 215.
Ohta, H. (1972). *Asahi Glass Co. Res. Lab. Rep.* **22**, 117.
Ohta, H. (1975). *Glass Technol.* **16**, 25.
Ohta, H. (1977). *J. Non-Crystall. Solids* **24**, 61.
Ohta, H., and Hara, M. (1970). *Asahi Glass Co. Res. Lab. Rep.* **20**, 15.
Olcott, J. S. (1968). U.S. Patent 3,395,998.
Olcott, J. S. and Stookey, S. D. (1962). *In* "Advances in Glass Technology" (Papers of *Int. Congr. Glass, 6th, Washington, D.C.*), pp. 400–403. Plenum Press, New York.
Partridge, G., and McMillan, P.W. (1974). *Glass Technol.* **15**, 127.
Pavelcheck, E. K., and Doremus, R. H. (1974). *J. Mater. Sci.* **9**, 1803.
Poole, J. P., and Snyder, H. C. (1975). *Glass Technol.* **16**, 109.
Plumet, E. (1970). U.S. Patent 3,505,047.
Plumet, E., and Toussaint, F. (1972). U.S. Patent 3,674,454.
Proctor, B. (1962). *Phys. Chem. Glasses* **3**, 7.
Proctor, B. A., Whitney, I., and Johnson, J. W. (1967). *Proc. R. Soc. London Ser. A* **297**, 534.
Rauscher, H. E. (1967). *Symp. Surfaces Glass and Its Mod. Treatments* pp. 45–54. Union Scientifique Continentale du Verre, Charleroi, Belgium.
Ray, N. H., Stacey, M. H., and Webster, S. J. (1967). *Phys. Chem. Glasses* **8**, 30.
Rinehart, D. W. (1966). British Patent 1,018,890.
Rinehart, D. W. (1967). U.S. Patent 3,357,876.
Rood, J. L. (1972). *In* "Introduction to Glass Science" (L. D. Pye *et al.*, eds.), pp. 373–389. Plenum Press, New York.
Rothmund, V., and Kornfeld, G. (1918). *Z. Anorg. Allg. Chem.* **103**, 129.
Schaeffer, H. A., and Heinze, R. (1974). *Glastech. Ber.* **47**, 199.
Schulze, G. (1913). *Ann. Phys.* **40**, 335.
Shonebarger, F. J. (1970). U.S. Patent 3,502, 454.
Shonebarger, F. J. (1971). U.S. Patent 3,615,319.
Spiegler, K. S., and Coryell, C. D. (1952). *J. Phys. Chem.* **56**, 106.
Spoor, W. J., and Burggraaf, A. J. (1966). *Phys. Chem. Glasses* **7**, 173.
Stern, K. H. (1966). *Chem. Rev.* **66**, 335.
Stern, K. H. (1972). *J. Phys. Chem. Ref. Data* **1**, 747.
Stookey, S. D. (1970). U.S. Patent 3,498,803.
Sundheim, B. R. (1964). "Fused Salts." McGraw-Hill, New York.
Symmers, C., Ward, J. B., and Sugarman, B. (1962). *Phys. Chem. Glasses* **3**, 76.
Terai, R. H. (1969). *Phys. Chem Glasses* **10**, 146.
Terai, R. H., and Hayami, R. (1975). *J. Non-Crystall. Solids* **18**, 217.
Tsekhomskii, V. A., Mazurin, O. V., and Evstropev, K. K. (1963). *Sov. Phys.-Solid State* **5**, 426.
Urnes, S. (1973). *J. Am. Ceram. Soc.* **56**, 514.

Varner, J. R. and Lang-Egelkraut, R. (1977). *In* "Non-Crystalline Solids" (G. H. Frischat, ed.), pp. 465–470. Trans. Tech. Publ., Aedermannsdorf, Switzerland.

Varshneya, A. K. (1975). *J. Am. Ceram. Soc.* **58**, 106.

Varshneya, A. K. and Mencik, Z. (1974). *J. Am. Ceram. Soc.* **57**, 170.

Varshneya, A. K., and Milberg, M. E. (1974). *J. Am. Ceram. Soc.* **57**, 165.

Varshneya, A. K., Cooper, A. R., and Cable, M. J. (1966). *J. Appl. Phys.* **37**, 2199.

Vassamillet, L. F., and Caldwell, V. E. (1969). *J. Appl. Phys.* **40**, 1637.

Ward, J. B., Sugarman, B. and Symmers, C. (1965). *Glass Technol.* **6**, 90.

Weber, N. (1965). U.S. Patent 3,218,220.

Zijlstra, A. L., and Burggraaf, A. J. (1968). *J. Non-Crystall. Solids* **1**, 49.

MATERIALS INDEX

271

SUBJECT INDEX